THE AMERICAN TORT PROCESS

THE
AMERICAN TORT PROCESS

John G. Fleming

CLARENDON PRESS · OXFORD
1988

Oxford University Press, Walton Street, Oxford OX2 6DP
Oxford New York Toronto
Delhi Bombay Calcutta Madras Karachi
Petaling Jaya Singapore Hong Kong Tokyo
Nairobi Dar es Salaam Cape Town
Melbourne Auckland
and associated companies in
Berlin Ibadan

Oxford is a trade mark of Oxford University Press

Published in the United States
by Oxford University Press USA

British Library Cataloguing in Publication Data
Fleming, John G. (John Gunther)
The American tort process.
1. Torts—United States
I. Title.
347.3063   KF1250
ISBN 0-19-825597-7

Library of Congress Cataloging in Publication Data
Fleming, John G.
The American tort process/John G. Fleming.
p. cm.   Includes index.
1. Torts—United States.   I. Title.
KF1250.F55 1988
346.7303—dc 19   [347.3063] 87-35214   CIP
ISBN 0-19-825597-7

Text Processed by The Oxford Text System
Printed in Great Britain
at the University Printing House, Oxford
by David Stanford
Printer to the University

PREFACE

This book has its origin in an annual lecture to explain the unique features of the American tort system to a masters' class of foreign students at Berkeley. Over the years this opportunity for taking stock fortified my conclusion that what makes American tort law so peculiarly different from that of other countries, those of the common law no less than the civil law, is not their substantive doctrinal content so much as the institutional framework in which it operates. The dish, so to speak, may consist of the same basic ingredients but is metamorphosed by the manner of its cooking.

Conventional textbooks of American tort law contain an array of rules which differ only in detail from those familiar to students elsewhere. Especially the English lawyer would find both the structure and content of the doctrine almost like his own, perhaps a little more idiosyncratic and more certainly richer in experience. But this impression would leave him entirely unprepared for the abrasive, prominent and controversial role played by the tort process in American public life. The energizing elements are found in the institutional framework in which the system operates and which give the American tort process its entirely unique flavour.

The purpose of this book is to explore the nature and effect of the most important of these institutional features, such as judicial activism, the jury system, the personal injury bar and its contingent fees, among others. Most of these have been studied in depth, and one incremental value this book may have is to introduce this literature to a wider public. Even those familiar with the American tort system may find some value in bringing together in the compass of a single book a discussion of matters that are otherwise apt to be considered in isolation. In that sense, the sum may be greater than its parts.

While focusing primarily on the infra-structure of tort law, my purpose has been not only to describe and criticize those institutions as discrete phenomena, but to assess also their impact on the application and development of substantive doctrine. Such, for instance, is the effect of the contingent fee on punitive damages, themselves an almost unique American institution at least in the role they play in personal injury litigation. By relating adjectival to substantive law, I may avoid the charge of producing a version of *Hamlet* without the Prince of Denmark.

But the choice of Tort *Process* rather than *Law* as the title of this book is intended as a message that the focus in the first instance is on institutional aspects of the tort *system* and only sequentially on substantive doctrine more or less directly affected. 'Process' is of course a term commonly used to comprise the whole gamut of processing claims from filing to judgment or settlement; but such a comprehensive survey would have exceeded the compass of this study and diverted attention from its real aim to identify the truly unique features of the American *tort* process. Thus the book is not organized along progressive phases of judicial procedure, but rather by the hierarchy of personae who participate in and give colour to it: appellate judges, trial judge and jury, trial lawyers and insurance adjusters. Within this framework, however, most aspects of processing tort claims receive some attention.

Several phenomena here dealt with, such as judicial activism and mass litigation, are by no means confined to the tort context; others, such as the jury system and the personal injury bar, are peculiarly tort related. My treatment of the former topics has been deliberately confined to their bearing on tort litigation, both for reasons of space and because of the abundant literature dealing with them in general terms. Since writing this text, the publication of *Form and Substance in Anglo-American Law* by Atiyah and Summers now offers an admirable study specifically comparing Anglo-American legal reasoning, theory and institutions from just such a general point of view.

Other disclaimers are also in order. While it is my thesis that the reality of the American tort system cannot be grasped by its doctrinal component alone, indeed that the institutional framework in which it operates has shaped many of its doctrines, this does not deny the moulding of other cultural and ideological forces on the changing face of legal doctrine.

G. Edward White's *Tort Law in America: An Intellectual History*
(1980) has mined that vein. Not that these two aspects are
really separable. For in an enterprise so close to the public
pulse as is the tort system in America, pervasive cultural
imperatives have obviously left their mark on one as on the
other. Thus judicial activism is at once an expression of (post-)
realist legal theory and the product of the great upsurge of
political liberalism released by the New Deal, whose deep
imprint on the tort system was brought about by judges bent
on social reform and by an aggressive exploitative plaintiffs'
bar. Moreover, tolerance of that activism is rooted more in
the prevailing belief in political pluralism (checks and balances)
than in any autochthonous legal theory. So also, the char-
acteristically pragmatic, no-nonsense, American philosophy of
life favoured a frankly instrumentalist view of the law's
function, just as its strong democratic ideology promoted a
judiciary armed with a social agenda and a classless, essentially
entrepreneurial legal profession.

   This book is not thematic in the sense of advocating specific
solutions or reforms. Critique there is aplenty but it is sceptical,
not missionary. The aim is a modest one of explaining
the workings of the system without entering the larger
contemporary debates about the scope and justification of
judicial law-making or the problematic survival of tort law.
Thus current academic fads, like economic analysis of law
(market theory) or deconstructionism (critical legal studies)
are not germane to my discourse, and speculation about the
ultimate fate of tort law is muted.

   This book none the less carries a clear message of the
'crisis' which in the view of many observers is besetting
the contemporary American tort system. Some blame the
insurance industry, its symbiotic companion. Others perceive
systemic flaws in the current direction of tort liability itself.
A powerful alliance has been impelling tort law from its
traditional commitment to inter-personal corrective justice to
a more ambitious task of social welfare. The liberal-radical
political constituency, which comprises also a large segment
of the legal establishment, has found an unexpected theoretical
ally in economic theorists who favour an expansive role for
tort liability as a means of deterring accident-prone conduct

and discouraging risky activities by raising their competitive costs. Social welfare can thus be maximized by matching no-fault compensation for victims with internalizing the cost to the risky enterprise. Opponents of this tendency divide among those who would stop this drift and return to the traditional notion of justice between man and man, and others who counsel abandonment of the whole tort system and its replacement by no-fault compensation with benefits proportioned to affordable costs. The contribution of the present study to this debate is to help explain how we got where we are, rather than whether we are going in the right direction or how to get off.

To end as I began on a personal note, the stimulus for putting these thoughts on paper was an invitation to Cambridge as the Goodhart Professor of Law for 1987-8. This book records in substance the lecture course I gave on that privileged occasion in memory of a great Anglo-American tort scholar who helped me, as so many others, in my first steps in that discipline.

Thanks are due also to several of my colleagues at Berkeley who critiqued portions of the manuscript, to Allan Ottaway and the editorial staff at Oxford University Press who saved me from some infelicities of style, to Gail Overstreet who patiently word-processed repeated drafts, and to Richard Hart of Oxford University Press who lent me his encouragement throughout this venture.

Berkeley and Cambridge

John G. Fleming

CONTENTS

THE TORT SYSTEM IN GENERAL PERSPECTIVE

In America tort liability is big business. More than 800,000 tort claims are filed each year and between 15,000 and 20,000 are tried, almost all before juries.[1] Multimillion-dollar awards, once headline-makers, have become commonplace. The first million-dollar award, a combination of compensatory and punitive damages, was rendered in 1962 in a defamation action;[2] nowadays there are more than 250 a year, many of them in products liability and medical malpractice suits.[3] Mass accidents and mass product disasters have repeatedly driven settlements into hundreds of millions of dollars, even billions, besides necessitating new procedures for administering such mass litigation.[4] In 1985 compensation to tort plaintiffs as the result of legal proceedings alone amounted to $20–5 bn.[5] One estimate put the total cost of the US tort system at $68 bn.[6]

[1] Kakalik and Pace, *Costs and Compensation Paid in Tort Litigation* 5–15 (1986); *State Court Case Load Statistics: Annual Report 1984*, table 14, 33 (1986). About 3.6 filings in courts of general jurisdiction per 1,000 population, or 1 in 275. (England recorded 50,000 writs for personal injury, or 1 in 1000: *Judicial Statistics, Annual Report* 1986.) Even so, only about 1 in 5 tort disputes result in litigation (Trubek *et al.*, 'The Costs of Ordinary Litigation', 31 *UCLA L. Rev.* 83 at 87 (1983)). The percentage of cases tried varies, from 3% in auto to 12% in medical malpractice cases (Kakalik and Pace at 80–1).

[2] *Faulk* v. *Aware, Inc.*, 231 NYS 270 ($1m. compensatory plus $25m. punitive damages for defamation).

[3] Frank, 'Multi-million Dollar Awards', *ABAJ*, Sept. 1984, p. 52. By mid-1984 1,118 million-dollar verdicts had been recorded.

[4] *Infra*, ch. 7.

[5] Kakalik and Pace, *supra*, at 67.

[6] Sturgis, *The Cost of the U.S. Tort System* (The Tillinghast Study) (1985). This figure does not include the costs of the court system or the litigants' time, calculated by Kakalik and Pace, *supra*, fig. 7.2, at 2% (court), 8–16% (litigants), depending on the type of case. This would raise the total above

The concentration on sensational verdicts in America may be misleading. Average awards tend to be high because of a relatively small number of very large awards. Median awards, however, are more modest because of the large proportion of motor vehicle accidents which dilute the impact of the mega-awards in product and medical malpractice actions.[7] All the same, the escalating trend of awards in 'extremely large' cases in recent years[8] is significant in profoundly affecting public perceptions about a current 'crisis' and its political fall-out.

A common theme among critics of the American scene is the diagnosis of a growing contentiousness in American life. They see Americans as 'rights-minded', possessed of an insatiable appetite for vindication, ready to take every dispute to litigation. 'Few Americans, it seems,' one of them observed, 'can tolerate more than five minutes of frustration without submitting to the temptation to sue.'[9] The huge number of lawyers seems to lend support to the belief that Americans are the most litigious nation in human history.[10]

$80bn. Kakalik and Pace, at 66, estimated the expenditure for tort litigation terminated in courts of general jurisdiction between $29 and $35bn. and compensation paid on other claims at $22.5bn. in 1985. These figures do not include costs of the insurance system, and are not therefore substantially lower than those of Tillinghast (at 75).

[7] The effect of mega-awards is seen in the contrast between average and median awards. In San Francisco average awards during 1975-9 were $179,000, median $26,000; during 1980-4 $302,000 and $62,000 (in 1984 dollars). See Peterson, *Civil Juries in the 1980s* 29 (1987). Median awards in England in 1974 were £1,819: *Report of Royal Commission on Civil Liability and Compensation for Personal Injury* (The Pearson Report), vol. ii, para. 86 (1978).

[8] Million-dollar awards in San Francisco were 5 in 1960-4, 21 in 1980-4, representing 14 and 58% of the total dollar awards, aggregating $7m. and $65m. respectively: Peterson, *supra*, at 33. There were comparable rises in Chicago and other California counties: ibid. The trend of higher awards, however, is general. Median awards trebled between the late 1970s and 1985: ibid., at 29.

[9] Auerbach, 'A Plague of Lawyers', *Harper's*, Oct. 1976, p. 37. The causes of this general phenomenon, apart from its specific bearing on the tort explosion, are beyond the scope of this study. Baxter, 'Behind the Legal Explosion', 27 *Stan. L. Rev.* 567 (1975) attributes it to loss of a common ethos and the insensitivity of regulators.

[10] While the proportion of lawyers in the population remained stable in the first half of the twentieth century, it increased sharply in succeeding

The steep rise of tort litigation is well documented and parallels, as we shall see, the spectacular rise of the plaintiffs' bar.[11] Indicative of the long-term trend is a study of five California counties in which tort filings rose from 1.2 per cent in 1903–4 to 12.4 per cent in 1976–7.[12] Even more spectacular has been the increase in the last twenty-five years, giving substance to the widespread perception of a 'tort explosion'.[13] For example, filings in federal courts almost doubled between 1961 and 1984;[14] in California state courts filings almost doubled between 1971 and 1984.[15] Paid claims and defence costs, it has been estimated, rose from 1.37 per cent of the gross national product in 1975 to 1.76 per cent in 1984, nearly

years. In 1960 there were about 286,000 lawyers, in 1983 622,000 (projected 1987: 750,000). The lawyer–population ratio in the US is therefore about 1 : 375, in England 1 : 1,650. For more comparative data see Galanter, 'Reading the Landscapes of Disputes: What We Know and Don't Know (and Think We Know) About Our Alleged Contentious, Litigious Society', 31 *UCLA L. Rev.* 4, 52–5 (1983), suggesting that the phenomenon is not unique to the United States and may be attributable to causes other than a peculiarly American neurosis, such as improved access to justice.

[11] Infra, ch. 5.

[12] Arthur Young & Co. & Public Sector Research, Inc., *An Empirical Study of the Judicial Role in Family and Commercial Disputes*, at table 10 (1981). This trend appears also from other studies: e.g. in Alameda County tort filings rose from 6% in 1890 to 27.1% in 1970: Friedman and Perceval, 'A Tale of Two Courts: Litigation in Alameda and San Benito Counties', 10 *L. & Soc. Rev.* 267, 281 (1976).

[13] The phenomenon is not confined to tort litigation. Those who speak of a 'litigation explosion' point to the fact that for several decades appellate case-loads have been doubling roughly every 10 years and trial court case-loads every 10 to 20 years. See Marvell, 'There *is* a Litigation Explosion', *Nat. L. J.*, 19 May 1986, who warns, however, against explaining this rise as due to a greater propensity to sue rather than as being related to an enormous expansion of the economy and other activities. Over the last decade tort litigation has remained steadily around 10% of all civil litigation: Kakalik and Pace, *supra*, at 10.

[14] From 21,205 to 37,522: *Annual Report of the Administrative Office of the U.S. Courts*.

[15] From 57,624 to 97,068: *Annual Reports of the Judicial Council of California*. Nation-wide, the current growth rate of filings has been estimated at 3.9% p.a. (i.e. doubling every 18 years): Kakalik and Pace, *supra*, at 13; Marvell, 'Civil and Tort Case Loads' (mimeo, 1986). The annual growth rate of compensation paid 1980–5 has been estimated at 12% for auto and 17% for other cases (cf. Consumer Price Index (CPI) 7%): Kakalik and Pace, at 16–36.

nine times as much as in 1950.[16] While the spectacular increase in tort claims in recent years has been the source of much consternation, it can be explained also in more sympathetic terms. Professor Galanter, for example,

does not view contemporary litigation as an eruption of pathological contentiousness, a dangerous and unprecedented loosening of needed restraints, or the breakdown of a common ethos or of community regulation. Instead, I see the contemporary pattern of disputing as an adaptive (but not necessarily optimal) response to a set of changing conditions. There have been great changes in the social production of injuries as a result of, among other things, the increased power and range of injury producing machinery and substances. There has been a great increase in social knowledge about the causation of injuries and of technologies for preventing them; there has been a wide dissemination of awareness of this knowledge to an increasingly educated public. There is an enhanced sense that harmful and confining conditions could be remedied. . . . In the light of all these changes, the pattern of use is conservative, departing relatively little from earlier patterns.[17]

Indeed, the tort system when viewed in the larger perspective of other social security arrangements merely shares with them a general, if spectacular, rise in American values of 'entitlement'.[18] In the period 1933–50 GNP, workers' compensation and tort costs experienced the same growth rate, increasing by a factor of five, but thereafter the deviation became very marked. In the following thirty-five years, when GNP rose only 2.9 times, workers' compensation rose by a factor of 7.2, health care 7.1, tort 8.7, government health 12.5, and social security 18.2.[19] This comparison suggests clearly

[16] The Tillinghast Study, *supra*. In recent years (since 1981) the annual growth rate has slowed to 3.9% (3% per capita): 8 *Nat. L.J.* no. 49, p. 38.

[17] Galanter, *supra*, at 69–70.

[18] 'Entitlement' has become a synonym for welfare rights. See generally Reich, 'The New Property', 73 *Yale L. J.* 733 (1964); id., 'Individual Rights and Social Welfare: The Emerging Social Issues', 74 *Yale L .J.* 1245 (1965); tenBroek and Wilson, 'Public Assistance and Social Insurance—A Normative Evaluation', 1 *UCLA L. Rev.* 237 (1954).

[19] These statistics are derived from the Tillinghast Study, *supra*. The figures are adjusted for inflation, which increased by a factor of 4.3. Medical costs in particular have risen steeply and disproportionately to the general cost of living reflected by the CPI. This is part of the general phenomenon of 'cost disease' afflicting personal services. But because these services are closely

that the tort phenomenon, far from being unique, is part of a general development in American society—a marked shift of resources to social welfare expenditures in response to heightened public expectations of entitlement.

Another parallel is that, just as the cost of social security has increasingly been drawn into public debate, so also have the cost (and benefits) of the tort system. Powerful vested interests are arraigned against each other: on one side those of the legal profession, principally plaintiffs' attorneys, who directly profit from inflated pay-outs of the system; on the other side, target defendants who have to pay the tally: the insurance and manufacturing industry, the medical profession, and public entities.

Whereas at one time, not so long ago, both plaintiff and defence interests were ardent supporters of the tort system, each seeking to manipulate it to its own advantage, more lately the defence side has increasingly lost confidence in its ability to 'live with it', let alone to control it as it did in days long gone by. Hence the withdrawal of liability insurers from certain states and lines of business or their demand for unaffordable rates, as in the case of the current medical and government insurance 'crisis'. These protests have been accompanied by belated campaigns to reform the tort system or replace it by alternatives, such as no-fault insurance. In sum, the very continuance of the tort system in its present form is for the first time entering the agenda of American public debate.

Central to that debate is, of course, an evaluation of the tort system's benefits and costs. I propose to address that problem briefly by reference to four aspects on which criticism has principally focused: (1) the explosion of tort law, (2) the transaction costs of the system, (3) the casualty insurance industry, and (4) reform proposals.

---

associated with the quality of life, it would be a mistake to judge their utility merely by their rise of opportunity costs. See Baumol and Oates, *The Theory of Environmental Policy*, ch. 16 (1975).

1. THE EXPLOSION OF TORT LAW

Until the mid-twentieth century the range of tort liability, though slowly expanding, remained under tight judicial controls. Although the remarkable breakout of liability for negligent harm in the nineteenth century had established the framework for an effective accident law, it was accompanied by control devices which ensured that the burden of liability would not exceed society's capacity to absorb it. Hence the invention of defences like contributory negligence, assumption of risk, and the fellow-servant rule, which until the advent of workers' compensation for all practical purposes precluded recovery for work injuries in the belief, no doubt, that industry was not ready to shoulder the burden. These defences, particularly that of contributory negligence, also disqualified a substantial proportion of other accident claims and heavily discounted settlement values. Similarly, responsibility for carelessness was circumscribed by the twin requirements of a 'duty of care' and 'proximate cause'. Both were policy controls with the specific function of limiting the burden of liability. Thus the 'floodgates'[20] remained closed against products liability, against liability for omissions, liability for mental disturbance, and for purely pecuniary loss. In addition, tight judicial controls at the trial and appellate level protected defendants from undue exposure to the vindictiveness of juries.

With few exceptions, judges were recruited from the ranks and supporters of the dominant class of society whose ideology and self interest favoured the cause of industrial and business enterprises, the target defendants (then as now) of tort claims. Some historians have gone to the length of concluding that 'the thrust of the rules, taken as a whole, approach [*sic*] the position that corporate enterprise would be flatly immune from actions sounding in tort'.[21] They speak of the law

---

[20] The *fons et origo* was Lord Abinger in *Winterbottom* v. *Wright* (1842) 10 M. & W. 109, 114. For a categorical modern repudiation see Lord Roskill in *Junior Books* v. *Veitchi Co.* [1983] 1 A C 520, 539.

[21] Friedman, *A History of American Law* 417 (1973). In similar vein, Horwitz, *The Transformation of American Law* 63–108 (1977), criticized for want of empirical proof: McClain, 68 *Calif. L. Rev.* 382, 394–5 (1980).

providing a subsidy to economic enterprise, of ruthlessly sacrificing the individual victim's right to compensation, even of the law as an 'engine of oppression'.[22] Others have also inculpated the later Langdellian school of legal theory and education which allegedly dehumanized legal rules in an effort to gain scientific respectability for them.[23] But it is one thing to say that nineteenth-century tort law, despite discarding strict liability in favour of 'no liability without fault', took a narrow, puritanical view of entitlement to compensation; it is another to accuse it of a deliberate policy to subordinate all else to a goal of economic expansion.[24] Indeed, a noted lawyer–economist has given his benediction to the overall record during this era for reflecting a design to bring about an efficient (cost-justified) level of accidents and safety rather than any systematic bias in favour of industrial growth.[25]

The brake on the expansive potential of tort liability was relaxed only gradually, in step with a more compassionate attitude towards the victims of accident and growing confidence in the ability of tort defendants to absorb these losses without unduly discouraging enterprise and initiative. The development and growth of liability insurance, initially designed to protect would-be defendants from ruinous liability, began to be viewed also as serving the purpose of compensating victims at a cost that would, by this device, be spread among larger segments of the public benefiting from the accident-causing activity. The remarkable contributions to tort law by Judge Benjamin Cardozo of the New York Court of Appeals (1914–32) well illustrate the prevailing view among fair-minded judges that they should proceed on a course of

[22] Ibid., at 410, 417.

[23] White, *Tort Law in America* ch. 2 (1980).

[24] See Schwartz, 'Tort Law and the Economy in 19th Century America: A Reinterpretation', 90 *Yale L. J.* 1717 (1981). His study of all nineteenth century California and New Hampshire Supreme Court tort cases did not support either the subsidy or Posner's efficiency hypotheses.

[25] Posner, 'A Theory of Negligence', 1 *J. Leg. Stud.* 29 (1972), which develops the general thesis that the negligence standard fixes the right balance for cost-justified investment in safety. This article is based on a statistical sample of appellate decisions between 1875 and 1905. Cf. Holmes's less theoretical theorem that strict liability unduly hampers initiative (*The Common Law*, lecture 3 (1881)).

cautious reform with due respect for history, equity, and concern for costs.[26] Thus fell the 'citadel of privity'[27] that had protected negligent manufacturers against injured consumers,[28] the blanket denial of liability for negligent misrepresentation[29] and to rescuers.[30] On the other hand, the continued immunity of accountants to larger groups of investors[31] and the strange restriction of liability to foreseeable plaintiffs (and perhaps consequences)[32] betrayed his conservative insistence on limits. It was permissible to explore the penumbra of accepted principles, to take a cautious step forward here and there, but not to break the bounds of the conventional structure. The validity of the fault-based negligence doctrine was not questioned, let alone challenged. Indeed, much of the appellate effort went into preserving a proper balance between judge and jury, with unquestioning faith in the ability of the legal process to maintain the values of the existing system of social ordering.

The explosion of tort law had to await the rapid transformation of American society after World War II. Its intellectual tap-root may have been the scepticism of the Realist School which, emanating from the Eastern law schools in the late 1920s, was spreading among the ranks of judges and lawyers and by the 1950s became the accepted gospel of opinion-makers inside and outside the classroom.[33] It purported to tear the veil of hypocrisy from the image of the judicial process by picturing it as essentially result-orientated rather than following the logical imperatives of doctrine, and by admonishing judges to a creative role in social engineering. Legal precedents should not be taken at face value for what

[26] See generally White, *supra*, ch. 4; Seavey, 'Mr. Justice Cardozo and the Law of Torts', 39 *Colum. L. Rev.* 20 (1939). Cardozo explained his judicial philosophy in *The Nature of the Judicial Process* (1921).

[27] *Ultramares Corp.* v. *Touche*, 255 NY 170, 180; 174 NE 441, 445 (1931).

[28] *McPherson* v. *Buick Motor Co.*, 217 NY 382, 111 NE 1050 (1916).

[29] *Glanzer* v. *Shepard*, 233 NY 236, 135 NE 275 (1922).

[30] *Wagner* v. *International Railway Co.*, 232 NY 176, 133 NE 437 (1921).

[31] *Supra*, n. 27.

[32] *Palsgraf* v. *Long Island Railroad Co.*, 248 NY 339, 162 NE 99 (1928).

[33] See Twining, *Karl Llewellyn and the Realist Movement* (1973); Summers, *Instrumentalism and Legal Theory* (1982); White, *supra*, ch. 3. A forerunner was Roscoe Pound's sociological jurisprudence.

they said, but for what they did. Behind the mask of value neutrality lurked the reality of bias and prejudice, social and economic. Law always had been a power play; the honest thing was not to deny this reality but consciously to use it for promoting desirable social ends.

This orientation legitimized judicial activism, the hallmark of the new era. No longer recognizing the legislature as the only legitimate law-making body, courts began to heed the call for reform, in response almost invariably to the mounting pressure from the plaintiffs' bar. From the Kennedy years onward civil rights activists were increasingly successful in manœuvring the courts into interventionist roles, restructuring central institutions by desegregating schools, reapportioning legislative districts, and humanizing prisons and public hospitals, by appealing to vague constitutional mandates or enforcing federal civil rights legislation against laggard bureaucracies.[34] The revolutionary feature in this development was that, instead of contenting themselves with the traditional negative function of constitutional adjudication in restricting the exercise of legislative and executive powers, the courts were embarking on restructuring political, social, and economic institutions.

Taking the cue from these primarily federal, judicial ventures, state courts also began to assume an overriding mandate actively to promote social reform. The general legislative inertia in state capitals gave a semblance of justification to the plea that the judicial branch was merely assisting, not usurping, the legislators' task of keeping the law abreast of changing social expectations. Besides, the 'American system' was built on checks and balances, on political as well as demographic and social pluralism, so that the notion of shared law-making did not strike such a discordant note. In truth, social activists were now looking to the courts for implementation of their agenda because judges proved in

[34] Institutional reform litigation originated in *Brown* v. *Board of Education* (*Brown II*), 349 US 294 (1955) which directed the district court to implement the right to non-segregated schooling established by *Brown I*, 347 US 483 (1954). See generally Horowitz, 'Decreeing Organizational Change: Judicial Supervision of Public Institutions', 1983 *Duke L. J.* 1265; id. *The Courts and Social Policy* (1977).

general to be more sympathetic to their aspirations than legislators. Tort reform in legislatures had to contend with the strong lobbies of the insurance industry and other defence interests, whereas courts were more amenable to arguments couched in terms of principle rather than of costs. Judges of more liberal political orientation than in the past, combined with the growing sophistication and muscle of the plaintiffs' bar, augured well for the cause of tort reform.

The process commenced on a relatively modest note by appeal to rationality rather than radical reform. Such was the technique of Justice Traynor in his famous concurring opinion in the *Escola* case in 1944.[35] Unlike the majority of the California Supreme Court, he spurned the device of *res ipsa loquitur* for upholding a plaintiff's verdict in an exploding bottle case in the absence of direct proof of fault, instead proposing outright strict liability. Doctrinal discontinuity could be masked by pointing to the already established application of strict (warranty) liability to food and drink, the present being merely a small incremental extension to containers thereof. Moreover, the proposal was not intended to affect the outcome of cases so much as to repudiate the pretence that consumer protection was still linked to the fault principle. The crucial steps in the reorientation of the law had been taken earlier when warranty was freed from the shackles of privity in cases of food and drink and *res ipsa* was being extended in consumer actions seemingly as an instrument of judicial policy rather than as a mere rule for evaluating circumstantial evidence. More volatile was Justice Traynor's policy argument that strict liability would promote a shifting of loss from the unwary consumer to the manufacturer who is best suited to afford protection against the constant risk of defective products. Even if not negligent, he was responsible for the product's reaching the market.[36]

[35] *Escola* v. *Coca-Cola Bottling Co.*, 24 Cal. 2d 453, 150 P. 2d 436 (1944). For more extended discussion of products liability see *infra*, ch. 2, at n. 106.

[36] Apart from his enthusiasm for strict products liability and abolition of the archaic immunities, Justice Traynor's record in the torts field was conservative. Thus he voted against strict liability for automobile drivers, against liability for mental disturbance by bystanders, against liability for an accident by a car thief (after leaving ignition key in unlocked car) and for

The seed thus planted took about twenty years to germinate. In the interval judicial tort reform was mainly preoccupied with abolishing charitable, family, and sovereign immunities.[37] This was accomplished by proclaiming a constitutional principle that judge-made rules could be undone by judges without legislation. Notably, none of these reforms involved a departure from basic doctrine. Justice Traynor's public policy argument, however, received a major boost from the eventual triumph of strict liability for all defective products in the early 1960s.[38] It lent credibility to a new policy argument that shifting of loss suffered by victims of accident could be justified on the ground that the defendant was in a strategic position, if not to prevent accidents, at least to provide compensation. The cost might act as an incentive to greater investment in safety and could in any event be spread by insurance or higher prices among the consumer public.

This theorem not only broke with the nineteenth-century commitment to fault liability; even more radical was its departure from the age-old view of tort liability as an instrument of corrective justice. For it lent respectability to the view that compensation could be justified by appeal to broader notions of social and economic policy, transcending interpersonal equities, like economic efficiency or redistribution of wealth on the pattern of social insurance. The use of tort law for such purposes, attractive alike to economists and social reformers, became the more plausible in the American context because of the dearth of public social security programmes which in other countries of the West have long furnished a safety net for the victims of misfortune. If the legislators were unwilling to impose a new direct tax burden on the public for such expenditures, the courts were prepared to convert

strict control of jury awards for pain and suffering: *Maloney* v. *Rath*, 69 Cal. 2d. 442, 445 P. 2d 513 (1968); *Amaya* v. *Home Ice, Fuel & Supply Co.*, 59 Cal. 2d 295, 379 P. 2d 513 (1963); *Richards* v. *Stanley*, 43 Cal. 2d 60, 271 P. 2d 23 (1954); *Seffert* v. *Los Angeles Transit Lines*, 56 Cal. 2d 498, 364 P. 2d 337 (1961). See generally, Malone, 'Contrasting Images of Tort: The Judicial Personality of Justice Traynor', 13 *Stan. L. Rev.* 779 (1961); White, *supra*, ch. 6.

[37] *Infra*, ch. 2.

[38] See Prosser, 'The Assault Upon the Citadel', 69 *Yale L. J.* 1099 (1960); id., 'The Fall of the Citadel', 50 *Minn. L. Rev.* 791 (1966).

tort law into a welfare system.[39] Although the overhead cost of that programme was daunting, the implicit tax on the consumer was disguised and, at least initially, passed without much protest to the general public. That judicial decisions often entail financial consequences beyond the immediate litigants is unavoidable and unexceptionable. But to raise redistribution of wealth to a primary goal of tort law was a radical departure from our cultural tradition.

The result has been what is widely hailed as an 'explosion' of tort liability. By no means confined to products liability, it is reflected in the postulate that 'where there is a wrong (? harm), there is a remedy'. Most pervasively it has stimulated the search for 'deep pockets' on behalf of claimants victimized by tortfeasors without assets.[40] The high percentage of un- or under-insured tortfeasors, even in motor-car accidents, ensures that the accident cost is in practice often met by defendants, such as municipal authorities responsible for street maintenance, whose share of fault may be negligible. In addition, it has encouraged a vast extension of 'duties of affirmative action' so as to impose responsibility on individuals and entities for mere omissions to prevent the risk of accident-prone behaviour by others; hence the liability visited on owners of commercial and residential premises for robberies and rapes, for failures by public officials, insurance agents, and the like to inspect the work performed by others, for the ubiquitous failures to warn consumers against misuse of products and an unimaginable array of dangers. In short, anyone even remotely related who could have interceded to prevent an accident is exposed to the risk of substantial liability.

In the eye of conservative critics the new ideology has taken the received tort law off its hinges. As stated at the beginning of this section, the development of the law had previously

[39] *Infra*, ch. 2.

[40] An extreme example is the 'successor liability' of *Ray* v. *Alad Corp.*, 19 Cal. 2d 22, 560 P. 2d 3 (1977) where Corporation I was held responsible for an injury by a defective product manufactured by Corp. II whose assets Corp. I had acquired for a fair price, after which Corp. II had dissolved. This novel form of vicarious liability the Court considered 'fair', because the plaintiff could no longer pursue the shareholders or officers of Corp. II. What if Corp. II had liquidated its assets piecemeal or its assets had fallen far short of the plaintiff's award? Whose insurance, if any, would cover this liability?

been restrained by concern that the cost of compensation would not excessively strain the resources of society. Without necessarily accepting the economists' teaching that the allocation of resources should here as elsewhere be guided by the sole criterion of efficiency,[41] the common law had been alive to the cost–benefit balance in deciding whether and how far to expand liability into new areas. Perhaps, in days gone by, the spectre of 'opening floodgates' was overdrawn; still, it revealed a keen sense of the long-term cost implications of a given decision. By contrast, the new ideology betrays little concern for the cost of its judicial welfare programme, in the insouciant or naïve belief that insurance and deep pockets will take care of the problem. The tort system has thus become about the only segment of the economy not subject to the discipline of prudent resource allocation: it is a programme without a budget.

The most sensational segments of tort litigation during this period have been products and medical liability. The vast expansion of liability and, in particular, the spectacular awards in these fields have received wide publicity, in turn encouraging ever more speculative claims and litigation. Concurrently, run of the mill motor-vehicle cases which dominated the torts scene in the preceding half-century, though still constituting the single largest category, have shrunk substantially in proportionate volume and size of awards.[42] The campaign for no-fault traffic compensation in the late 1960s met with only indifferent success and quickly exhausted the impetus for reform in that area, partly due to the disarray in the insurance industry.[43] By contrast, in medical and products liability the insurers at last found an issue on which they could form a

[41] The most prominent spokesman for that approach is Posner. See 'A Theory of Negligence', *supra*; *Economic Analysis of Law* ch. 10 (2nd edn. 1979).

[42] e.g., in California the percentage of motor vehicle cases filed fell from 72% to 57% between 1971 and 1984. *Report of Judicial Council of California 1985*, 106. In San Francisco the percentage of cases tried fell from 53% in 1960–4 to 38% in 1975–9. Also, while mean awards in malpractice and products cases rose by 500% and 300% and median by 150% and 300%, motor vehicle awards rose by only 160% and 73% respectively. Shanley and Peterson, *Comparative Justice: Civil Jury Verdicts in San Francisco and Cook Counties, 1959–1980*, 21 (1983).

[43] *Infra*, ch. 5, at n. 75.

united front, attract powerful institutional allies, and appeal for popular support. In the hyperbolic rhetoric of the proponents of reform, America was struck by a series of crises, the 'medical insurance', the 'products liability', and now compendiously the 'tort' crisis. The issue became politicized as the threatened defence interests attempted to mobilize public support in favour of legislative intervention, in opposition to the organized trial bar which rose in defence of its forensic gains. Thus, in a manner strange to foreign observers, tort liability entered the social and political debates of our time, less controversial only than abortion or the death penalty.

1.1   Products liability

Products liability attained its prominence after the adoption of strict liability in the 1960s and more particularly after its extension in the following decade from construction to design defects (including failure to warn) and the resort to punitive damages.[44] Filings in the federal courts alone rose from about 4,000 in 1976 to over 10,000 in 1983.[45] In Chicago jury trials rose from 3 per cent in the 1960s to 8 per cent in the 1970s.[46] Most spectacular was the explosion of awards. Nation-wide, the average product liability award rose from $340,000 in 1974 to $1.07m. in 1984.[47] The rise of jury awards in San Francisco and Chicago is indicated in Fig. 1.1. This trend has accelerated. During 1980–4 median awards were $200,000 in San Francisco, $187,000 in Cook County; average awards $1.05m. and $828,000. In California five awards in 1984 exceeded $3m.; in 1983 three awards totalled more than $20m.[48] Total awards have increased from $700m. in 1980 to about $1.8bn. in 1985.[49]

[44] Construction defects are isolated events, design defects impugn a whole production line. This increases exposure to liability multifold. See also *infra*, ch. 2, at n. 108.

[45] Of these in FY (Financial Year) 1983–4 2,800 were asbestos claims. See *Annual Reports of the Administrative Office of the U.S. Courts.*

[46] Peterson, *Compensation of Injuries: Civil Jury Trials in Cook County* 44 (1984); Peterson and Priest, *The Civil Jury: Trends in Trials and Verdicts, Cook County, Illinois* 42–4 (1982).

[47] Johnson and Higgins, *Risk and Insurance Management Services* (1985). Peterson, *supra*, at 55.

[48] *Nat. L. J.*, 25 Feb. 1986, p. 32.   [49] *The Economist*, 10 Jan. 1987, p. 51.

The pharmaceutical industry has been a special target of damage claims, with undesirable deterrent effects in general on the marketing of drugs and products (like contraceptive devices) driving at least one manufacturer into bankruptcy.[50] The swine flu epidemic in 1976 necessitated a government assumption of liability in order to induce the co-operation of vaccine manufacturers,[51] and in 1986 Congress eventually created a fund for all vaccine injuries.[52] By comparison, in Britain, there had not been (by 1987) a single successful lawsuit against a drug manufacturer.[53]

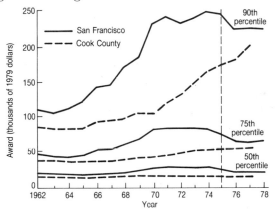

Source: Shanley and Peterson, *Comparative Justice: Civil Jury Verdicts in San Francisco and Cook Counties, 1959–1980* 55 (1983).

[50] A. H. Robins, the manufacturer of the Dalkon Shield. Contraceptive devices, including the Pill, have become increasingly unavailable. Huber, 'Safety and the Second Best: The Hazards of Public Risk Management in the Courts', 85 *Colum. L. Rev.* 277 (1985) mounts a powerful argument against short-sighted interdiction of innovative technology as in the long run counterproductive to safety.

[51] See Franklin and Mais, 'Tort Law and Mass Immunization Programs: Lessons from the Polio and Flu Episodes', 65 *Calif. L. Rev.* 754 (1977).

[52] National Childhood Vaccine Injury Act of 1986. As an alternative to the benefits of the compensation programme a plaintiff may bring a tort action, but the conditions of liability are circumscribed (e.g. warning to physician sufficient; no liability for unavoidable side-effects, with warning. In the last respect, latest case-law suggests the same result: see *Collins* v. *Karoll*, 186 Cal. App. 3d 1194, 231 Cal. Rptr. 396 (1986).

[53] For the Thalidomide affair see Insight Team of *Sunday Times* of London, *Suffer the Children: The Story of Thalidomide* (1979). A multi-plaintiff action against Eli Lilly's anti-arthritis drug Opren was settled: 1987 *N.L.J.* 1183.

The mounting tally of losses by the insurance companies inevitably led to a steep rise in insurance rates. In consequence, the President's Economic Policy Board in 1975 appointed an Interagency Task Force on Products Liability which submitted, in addition to a comprehensive report, the draft of a Model Products Liability Act.[54] This measure would have rolled back design defect liability to a negligence standard, strengthened judicial controls over juries, and introduced a ten-year statute of repose. Some of its provisions have been variously adopted by conservative state legislatures, but opposition to reform by consumer groups and the trial attorneys has stultified success elsewhere.[55] The scene then shifted to Washington where the Reagan administration eventually gave its blessing to national legislation despite its general state rights ideology. A federal version of the Model Act died in the US Senate in 1985,[56] but a new initiative was launched in 1986, the Danforth Bill, which would allow an 'expedited claim' against the manufacturer of a defective product for the net economic loss as an alternative to traditional proceedings for full tort damages.[57]

1.2 Medical malpractice

Aside from products liability, no other segment of tort litigation has experienced such rapid growth in the last twenty years as have claims for medical malpractice. The percentage of physicians sued nearly tripled in the period between 1978 and 1983. In 1983 there were 16 malpractice claims for every 100 doctors, compared with 1 for 65 in 1956; 60 per cent of gynaecologists reported having been sued, 20 per cent three or more times. Moreover, awards have sky-rocketed: the

[54] The Uniform Product Liability Act, published by the US Department of Commerce; 44 *Federal Register* 62714 (1979).

[55] See Vandall, 'Undermining Torts' Policies: Products Liability Legislation', 30 *Am. U. L. Rev.* 673 (1981); McGovern, 'The Variety, Policy and Constitutionality of Products Liability Statutes of Repose', ibid. 579. These legislative initiatives were more successful in the latest push 1985-7 (see *infra*, n. 85).

[56] The original Kasten bill (S. 100, 99th Congress), launched in 1982. A new version, emanating from the White House, was launched in May 1986: The Product Liability Reform Act of 1986 (Kasten Amendment No. 1814).

[57] S. 1999, 99th Congress.

number of million-dollar verdicts (1,963) doubled between 1972 and 1983, the average award in 1985 being $950,000. So did settlements: in 1982 more than 250 exceeded $1m., a tenfold increase in four years.[58]

Many causes have contributed to this phenomenon: more liberal rules of substantive and procedural law, a superior trial bar, more sympathetic juries impressed by the staggering cost of necessary medical care, and, not least, the public expectation of magical medical cure.[59] Although the success rate of jury trials lies well below average,[60] the risk of litigation is worth taking in view of the size of awards. The inevitable response has been for insurance companies to raise premiums in order to match these costs or to withdraw altogether from this line of business. In 1986 the estimated cost of medical liability insurance premiums reached a total of $4.7bn., up from $1.7bn. in 1983.[61] In some fields of medicine, like gynaecology and neurosurgery, six digit premiums have become standard.[62]

[58] See 38 *Okla. L. Rev.*196, nn. 2, 3 (1985), citing AMA Special Task Force on Professional Liability and Insurance in the 80s, *Report No. 1* (1984); *NY Times*, 4 Feb. 1985, 'Again the Malpractice Crunch'; Frank, *supra*. According to Jury Verdict Research average awards have risen from $300,000 in 1975 to $700,000 in 1984 and an estimated $1m. in 1985. Viewed over a longer period, the increase is equally remarkable: San Francisco jury awards rose from a median of $64,000 (average $125,000) in 1960–4 to $156,000 (average $1.179m.) in 1980–4: Peterson, *supra*, n. 7, at 22. See generally Danzon, *Medical Malpractice: Theory, Evidence and Public Policy* (1985), with special focus on economic analysis of data 1975–8.

[59] Although patients sue doctors much more frequently than 30 years ago, Danzon (*supra*, at 29) found that patients are not excessively litigious. Mid-1970 California studies suggest that no more than 10% of hospitalized patients injured by staff negligence file claims. Despite this degree of under-deterrence, Danzon argues for limitation on non-pecuniary damages and shorter periods of limitation.

[60] While the percentage of plaintiff victories in all jury trials in San Francisco and Chicago, 1959–80, was 59% and 52%, that in medical malpractice cases was only 35% and 33% respectively. Shanley and Peterson, *supra*, n. 42, at 11. Two surveys from insurance files, 1974–6, presented a similar picture: 43% of claims were abandoned, 50% settled, of the remaining 7% 1 in 4 succeeded: Danzon, *supra*, at 32.

[61] Government Accounting Office (GAO), *Report to U.S. Congress*, Sept. 1986.

[62] Obstetricians have become special targets of blame for any birth defects or accidents. Some have declined to accept new patients and are moving into less vulnerable specialities. According to the American College of Ob-

The insurance industry and health care organizations therefore launched a legislative campaign in the mid-1970s to curb liability, which succeeded in many states over the head of the plaintiff attorneys' lobby and, as we shall see, withstood most subsequent constitutional challenges.[63] These reforms appear to have achieved some reduction in the frequency and severity of claims[64] (though without retarding the continuing rise of insurance premiums),[65] and have provided a model for, and encouraged, reform proposals of general application beyond the medical context.

2. TRANSACTION COSTS

The most negative feature of the tort system is its staggering overhead cost. Compared with other accident compensation systems, even those administered by private insurance like American workers' compensation (30 per cent) and health insurance (15 per cent), let alone with state insurance funds like New Zealand's accident compensation plan (8 per cent),[66] its cost inefficiency is difficult to justify by any competing advantages over its competitors.[67] Even in England the

stetricians and Gynaecologists (ACOG) 12% had given up practice by 1985. Others proclaim on bumper stickers: LET THE LAWYERS DELIVER THE BABIES. *Time*, 24 Feb. 1986, p. 60.

[63] See Robinson, 'The Medical Malpractice Crisis of the 1970's: A Retrospective', 49(2) *Law and Contemp. Prob.* 277 (1986). *Infra*, ch. 3, at n. 37.

[64] Danzon, *New Evidence on the Frequency and Severity of Medical Malpractice Claims* (1966) estimated a reduction of 11–18% for offset of collateral benefits, 23% for caps on awards, and 8% for each one year reduction of the statute of limitation. None the less, astronomical awards still abound: e.g. $8.4m. for severe brain damage at birth (*Nat. L. J.*, 5 May 1986, p. 16).

[65] Premiums soared 100% in just 2 years (cf. with the 8% rise of CPI and 13% of Medical Care Index). See GAO study, *supra*. In 1984 physicians on average paid 4.2% of their income in premiums.

But in New York, pursuant to a mandated analysis of the insurance market under the Reform Act 1985, the Superintendent ordered a 15% reduction. See also *infra*, n. 84.

[66] The American figures are based on the Tillinghast study, *supra*, at 18; 'Social Security Programs in the United States', 49 *Soc. Sec. Bull.* 1 (1986). For New Zealand see Hodge, 'No-Fault in New Zealand: It Works', 50 *Ins. Couns. J.* 222, 230 (1983).

[67] See Fleming, 'Is There a Future for Torts?', 58 *ALJ* 131 (1984), 44 *La. L. Rev.* 1193 (1984).

Pearson Commission estimated that it cost 85p to deliver £1 in net benefits to the victim.[68] Studies in the United States raise operating costs to $1.07 for automobile[69] and $1.25 for product liability.[70] In the protracted asbestos litigation it has cost $1.59 in combined litigation expenses to deliver $1 to the average plaintiff.[71] As a result, in combination with the high component of damages for non-pecuniary injury, only about 15 per cent of the cost of the tort system accounts for out-of-pocket losses.

The high transaction costs are inherent in the system itself. First, in the nature of tort liability; second, in third-party insurance as a conduit for spreading the cost. The primary cause of the first is the adversarial relationship between claimant and compensation source. Liability to compensate is dependent on issues of fault and causation which require investigation and are frequently contested. The assessment of damages, tailored to each case, invites additional and usually even keener controversy. In sum, the system is geared to individual processing and does not favour economies of scale.

Legal fees constitute the principal component in accounting for the substantially higher American costs compared with the British. Not only are legal services generally more costly, but the tort plaintiff's attorney's contingent fee consumes from one-third to one half of the award, a proportion in many cases very much higher than taxed costs awarded to a successful English claimant.[72] In the result, legal costs incurred by both parties alone statistically amount to more than the benefit to the victim.[73] It should be remembered also that defendants

[68] *Pearson Report*, vol. i, *supra*, n. 7, para. 261. Otherwise expressed, operation costs aggregate to 45 cents of the premium dollar.

[69] Department of Transportation, *Motor Vehicle Crash Losses and Their Compensation in the U.S.* 47–52 (1971).

[70] *Interagency Task Force on Products Liability, Final Report*, ch. 5, 23–5 (1976). Kakalik and Pace, *supra*, at 70, estimate net pay-out to plaintiffs in all tort litigation at 46% of total litigation cost. For motor vehicle cases plaintiffs' share was larger (52% vs. 43%) because of larger defence costs: ibid., at xiii.

[71] Defence costs amounting to $0.95, plaintiff's $0.64. Otherwise expressed, the plaintiff's net award equalled 39% of the total cost. Kakalik *et al.*, *Variation in Asbestos Litigation Compensation Expenses* 86–91 (1984).

[72] *Infra*, ch. 6.

[73] A study by the ISO (Insurance Services Office) estimated that 38.5% of average settlements went to plaintiffs, 24.7% to their attorneys, and 36.6%

incur legal costs in processing *all* claims, not only those that are eventually successful, since American law does not countenance fee-shifting in favour of successful defendants any more than in favour of successful plaintiffs.

Next, substantial administrative costs are incurred as a result of employing private insurance for processing and paying claims. In the United States, insurance (even compulsory insurance) is obtained mostly through independent agents or brokers who take the top 10 or more per cent of the premium dollar. In turn, the insurance carrier itself incurs administrative expenses (investigation, legal defence, settlement) in processing claims. These so-called loss adjustment expenses amount to approximately 20 per cent of the premium dollar. Assuming an optimal target ratio for loss and loss adjustment of 60 per cent, this would leave 40 per cent for underwriting expenses, overhead allocations, commissions for agents and brokers, and profits.[74] But, as will be explained below, in several lines of business, insurance companies have in fact suffered deficits over the years which will either have to be recouped by increasing premiums or written off.

Finally, the expenses of the judicial system are incurred by government from filing claims to disposition. Of a total estimated expenditure for processing all civil cases in the United States, amounting to over \$2bn. in 1982, \$425m. was spent on tort claims: \$269m. on 661,000 state court actions filed, \$56m. on over 32,000 federal cases. This worked out as an average expenditure of \$407 per case in state courts and \$1,040 in federal courts.[75] In California the cost per case was \$511, per jury trial \$8,000.[76]

---

for defence fees. For cases proceeding to trial, the figures were 37.1%, 30%, and 32.9% respectively. Kakalik and Pace, *supra*, at 71, estimate net compensation in auto torts at 52%, in other torts at 43%, legal fees absorbing 37% and 38% respectively. See also Interagency Task Force, *supra*, ch. 5. The Minogue Report, para. 8.15 (1978) estimated that legal costs in Victoria (Austral.) 'can exceed 20% of the total payouts'.

[74] *Supra*, n. 70.

[75] Kakalik and Ross, *Costs of the Civil Justice System: Court Expenditures for Various Types of Civil Cases* 81–5 (1983).

[76] Kakalik, Eisenshtat, and Robyn, *Costs of the Civil Justice System: Court Expenditures for Processing Tort Cases* 63 (1982).

The low efficiency of the tort system suggests to its detractors that 'the real question is whether the American public will much longer tolerate a $68 billion system that returns 25 cents on the dollar to those who win the fault lottery, and nothing to those who lose'.[77]

3. THE INSURANCE INDUSTRY

Tort law and liability insurance enjoy a symbiotic relationship. Neither could exist without the other: without exposure to liability, insurance would not be needed; without insurance, tort liability would be an empty gesture, reducing the tort system to a negligible role of accident compensation and depriving target defendants of needed protection against financial catastrophe. The dual role of liability insurance in protecting both plaintiffs and defendants has been long recognized and its bearing on the public interest acknowledged in legislation mandating insurance for many activities, from motor traffic to nuclear facilities, in regulating terms of such insurance, and even in setting premium rates.

Escalating premiums have cast the insurance industry into the vortex of contemporary controversy over the future of tort law. (In 1984 Americans paid $76bn. in premiums to domestic non-captive casualty insurance companies.)[78] The insurance companies are committed to keeping the lid on liability both for the sake of their own profits and so that the cost of premiums will not price their product out of the market. Representing defendants, as they do, it has been their role to oppose plaintiffs' claims in court and in settlement negotiation; also to lobby for legislation limiting the incidence of liability. Their legislative targets have been to eliminate 'moral hazards',

---

[77] The Tillinghast Study, *supra*, at 19. Not included in the preceding tally of the tort system's costs are non-administrative costs, such as the cost of 'defensive medicine', estimated by one study of the AMA at $40bn. p.a.: *1985–86 Property Casualty Fact Book* (Ins. Info. Institute 1985). Against this would have to be set the cost of accidents prevented by the would-be deterrent effect of the tort system, which Danzon (*supra*, n. 58), for one, considered to be optimal.

[78] A. M. Best, *Averages and Aggregates, 1985*.

like collusion among family members and friends ('guest statutes')[79] and, more lately, to limit awards. The medical insurance crisis of the mid-1970s and the later general insurance crisis of the mid-1980s have both been evoked by the insurance industry to enlist public support by premium-payers for a major campaign to reform the tort system, including limitations on non-pecuniary losses and contingent fees, abolition of punitive damages, and provision for periodic payments in lieu of lump sums in catastrophic cases.[80]

For the insurance industry the villain of the piece is, of course, the trial bar. It is the contingent fees that, in its perspective, are the driving engine of ever escalating verdicts and settlements, aided by the misguided generosity of juries and the connivance of liberal judges. The plaintiffs' bar, for its part, points the accusing finger at the insurance industry for its hypocrisy in blaming others for the 'crisis' for which its own cupidity and incompetence alone are to blame. According to this scenario,[81] the insurers brought the trouble upon themselves by greedily writing insurance at unrealistic rates in order to take advantage of the high investment opportunities in the early 1980s and are now reaping the harvest of their imprudence. Besides, even if the casualty insurance industry had underwriting losses of $28bn. in the 1975–85 decade, it earned $100bn. from investments: a net profit of $72bn. or 7.2 bn. per year.[82] A more plausible explanation of their self-styled crises was a conspiratorial effort to reverse the wholesome expansion of legal protection against corporate and other malefactors for which the trial lawyers had fought so hard on behalf of the victimized public. The spectacular rise of premiums and even insurers' withdrawal from certain lines of business were just a ploy in the campaign to mobilize public

---

[79] *Infra*, ch. 3, at n. 12.

[80] These statutes and their constitutional challenges are discussed in ch. 3.

[81] See e.g. President Hinton of the CTLA, 'Protecting the Innocent Victim: Joint and Several Liability', *CTLA Forum*, vol. xvi no. 1, p. 7 (1986). Cf. Danzon, *supra*, n. 58, ch 6.

[82] GAO, *Report to U.S. Congress*, Aug. 1985. It is also pointed out that insurance stocks improved 30% in 1985, twice as much as the market as a whole.

opinion against tort liability and its guardians, the trial bar.[83] Reform should be aimed not at tort law but at closer supervision of insurance companies and their rating policies.[84]

All of these charges and countercharges contain some truths. In comparison with other branches of casualty insurance (let alone life assurance), liability insurance presents acute problems of calculability. Assessment of risk is peculiarly dependent upon accurate information about exposure. Rapid changes in the legal climate, unless closely monitored and quickly responded to, can lead to substantial deficits. But reaction by the insurance industry to the well-documented rise in awards during the past two decades was belated, and, when it occurred, seemed to be the more dramatic because of an understandable effort to recoup past losses. Significantly, the 'crises' were confined to the volatile areas of liability—products and medical malpractice—whereas in traditional areas, like traffic accidents, awards and premiums remained stable. In the case of products liability, the problem of calculability is moreover magnified by the mass exposure resulting from any single new product, which makes the risk difficult to identify in advance and spread economically.

The insurance industry's travail was aggravated by the contemporaneous volatility of the investment market. At times of rising interest rates, insurance companies reduce premiums, partly because they are hungry for cash to take advantage of more favourable investment opportunities, partly in response to consequential competitive pressures. Conversely, when interest rates fall, premiums rise. Precisely this sequence of events occurred: lower premiums during the high inflation of

---

[83] Investigation into conspiratorial fixing of premiums in violation of the anti-trust laws was abandoned by the Justice Department in May 1986 for want of sufficient evidence.

[84] This has borne fruit in a few states. W. Va. coupled a $1m. cap on non-economic damages with a mandate to insurers to open their books and provide an explicit explanation of how they derive their rates. *Wall St. J.*, 15 Apr. 1986, p. 6. Florida prescribed rate standards (based on rate of return) and distributions of excessive profits to policyholders: Tort Reform and Insurance Act 1986. Government rate setting or control over rates of liability insurance is virtually unknown. Supervision is generally aimed at ensuring that rates are adequate and not discriminatory, reasonableness being left to competition. See Keeton, *Basic Text on Insurance Law* 543–57 (1971).

1976–84, followed by a 70 per cent rise in premiums under steeply falling interest rates. And while insurance-related losses had been balanced by investment income, commercial liability and medical malpractice insurance produced 23.8 per cent of the property-and-casualty sector's $25.2bn. underwriting loss.

Tort reform statutes in thirty-nine states have effected modest changes of substantive and remedial law since the mid-70s,[85] most of them applicable only to medical malpractice and contemplating a short lifespan ('sunset laws'). Although promoted by the insurance industry, they have not resulted in any noticeable reduction of premiums, thereby fuelling the accusation that the reform campaign was just a disingenuous attack on the deserved gains of tort victims.[86] In truth, there are a number of good reasons why limitations on damages are unlikely to make a major dent on insurance costs.[87] Rather, their chief hoped-for effect was to send a legislative signal to the judges to reconsider the direction and pace of legal change.

Critics of the insurance industry may be right in arguing that the existing tort system provides no more than adequate compensation for the injured and serves as an efficient deterrent against pervasive safety failures by industry and professionals. But insurance is a profit-motivated business and cannot be blamed for withdrawing from unprofitable markets or adjusting its premiums to rising risks. The unasked question is whether society is prepared to foot the bill.

4.   LEGISLATIVE REFORMS

Legislative intervention has been episodic and of modest scope. The torch-bearers of reform have been the courts, not the legislatures. In general, the legislators acquiesced in the

[85] Following the first wave, 19 statutes were passed in 1986. See documentation in Donaldson, Hensen, and Jordan, 'Jurisdictional Survey of Tort Provisions of Washington's 1986 Tort Reform Act', 22 *Gonz. L. Rev.* 47 (1986–7). More followed in 1987.

[86] See *supra*, n. 84.

[87] The most important being that the reforms benefit defendants more than their insurers (typical coverage does not exceed 'caps'; punitive damages are not insurable).

exercise and programme of judicial activism, for reasons to be explored in the next two chapters. Only two exceptions should be noted here. The judicial abrogation of sovereign immunity trenched directly on the fiscal concerns of state and local government, which not infrequently countered by modifying its impact. Also, in a dwindling number of conservative jurisdictions adhering to the traditional allocation of constitutional powers, the more drastic reforms, like the adoption of comparative fault, were carried out by legislation.

For the most part, however, legislative intervention has been not to expand but to limit liability, to counteract the excesses of judicial activism. The promoters were invariably the casualty insurance companies, in alliance with the medical profession, manufacturing industry, or local government, depending on the issue. Two waves of counter-attack have so far occurred: the first 'medical crisis' of the mid-1970s which resulted in statutory limits on non-economic damages and contingent fees in a number of jurisdictions threatened by a withdrawal of insurers, and the second, more general 'tort crisis' of the mid-1980s aimed at generalizing these restrictions for all tort claims as well as modifying the 'joint and several liability' (deep pocket) rule.

Preceding these two crises was the campaign in the latter part of the 1960s to introduce a new system of first-party no-fault automobile insurance, which also drew its impetus from rapidly rising liability insurance premiums.[88] Most statutes that were finally enacted, however, fell far short of the Keeton–O'Connell model which aimed at substantially limiting recourse to tort liability. Only New York and Michigan enacted plans which provide a serious alternative to tort liability as a system of accident compensation. Otherwise, statutory no-fault compensation plans have not made as much progress in the United States as in other countries. Workers' compensation made a tardy entry in the second decade of the twentieth century after encountering prolonged constitutional challenge, but offers only modest benefits in comparison with tort damages and has never been upgraded into a social

---

[88] *Infra*, ch. 5, at n. 75.

security system as in most countries of Western Europe.[89] And while workers' compensation is the exclusive remedy against the employer,[90] tort claims are being pursued with increasing success against third parties, especially equipment manufacturers,[91] through widening loopholes even against the employer himself.[92]

By contrast, in most other developed countries insurance plans have made considerable advance in replacing tort liability as a source of accident compensation.[93] Since World War II national systems of social insurance provide free health care and basic monetary benefits for persons disabled by accident, sickness, even unemployment. Usually basic benefits are focused on need rather than income replacement. Job injuries have mostly retained the historical link with workers' compensation in providing income-related benefits, though subject to 'caps'. Since tort damages would be reduced by these benefits, resort to tort recovery by the victim is largely

[89] For a brief overview see Social Security Programs, *supra*, n. 66, at 28–37; Franklin and Rabin, *Tort Law and Alternatives* 682–703 (4th edn. 1987). In 1983 for a premium cost of $25bn. benefits of $17.5bn. covered over 13,000 deaths and 2.3m. injuries.

[90] Larson, *Workmen's Compensation Law*, vol. 2A, ch. 14 (1961-).

[91] According to an ISO study reported in *Final Report, Interagency Task Force on Product Liability* VII–85 (1978), 11% of product liability claims arose from work injuries but accounted for almost 50% of the total insurance payouts. This is confirmed by the Rand studies of jury awards in San Francisco where median and average awards for 'worker injuries' were more than twice the total and only surpassed by professional malpractice: Shanley and Peterson *supra*, at note 42. Significantly, this makes available to the victims two sources of no-fault compensation, a bonanza that motivated a statutory amendment in the parallel situation of claims for unseaworthiness by waterside workers. (The amendment confining claims to negligence was enacted in the Longshoremen and Harbor Workers' Compensation Act Amendments 1972.)

The workers' compensation insurer's right of subrogation against third-party tortfeasors, which in most jurisdictions is independent of fault on the part of the employer, raises additional concerns which have been addressed in reform proposals: see Franklin and Rabin, *supra*, at 699–701.

[92] Besides serious misconduct, the 'dual capacity' judicial gloss permits tort actions against the employer, e.g. for failing to notify an employee of the result of adverse medical tests or where the employee is injured by a product he is distributing to the public for his employer. See *D'Angona* v. *County of Los Angeles*, 27 Cal. 3d 661, 613 P. 2d 238 (1980).

[93] See e.g. Weyers, *Unfallschäden* (1971).

discouraged except in cases of severe injury when it might be worth his while to pursue damages for non-economic loss or, in the case of high income earners, for the excess income loss. Therefore the higher the scale of benefits and the more modest the prevailing tariff for non-pecuniary tort damages, the more tort liability is elbowed out by social insurance. This result is especially evident in Scandinavian countries,[94] as also in Britain where the social security system is now recognized as the 'senior partner' to tort as a source of accident compensation.[95] Moreover, in the countries just mentioned social security is also denied any right of recoupment from tortfeasors and their insurers so that, in contrast to the more traditional rule in France and Germany, social security remains the primary source of compensation.[96]

In addition to this safety net, various special compensation plans further reduce the role of conventional tort liability. The principal example is no-fault traffic accident compensation which, if it does not become the exclusive remedy (as in Quebec and Israel), takes care of an overwhelming proportion of such personal injuries.[97] Other special plans, such as for drug injuries or medical mishaps, provide additional disincentives to tort litigation, all the more so where, as in Sweden, the benefits are deliberately designed to cover the difference between general social security benefits and tort damages.[98] Finally, in New Zealand a comprehensive accident compensation scheme for all personal injury and death has replaced tort liability completely.[99]

[94] See Hellner, 'Compensation for Personal Injury: The Swedish Alternative', 34 *Am. J. Comp. L.* 613 (1986).
[95] *Pearson Report, supra*, vol. i, para. 1732 (1978). The annual cost of social security was then £421m., that of torts £202m.: ibid. 13. Writs in the Queen's Bench for tort claims in 1984 numbered 32,000, in the County Courts (up to £5,000) 27,000. In California, with half the population, tort filings numbered 100,000, i.e. about three times per capita.
[96] See Fleming, XI/2 *Int. Encycl. Comp. L.* ch. 11 §§ 18–62.
[97] See Tunc, XI/2 *Int. Encycl. Comp. L.* ch. 14.
[98] *Supra*, n. 2.
[99] See Palmer, *Compensation for Incapacity* (1979); id., 'Compensation for Personal Injury: A Requiem for the Common Law in New Zealand', 21 *Am. J. Comp. L.* 1 (1973).

What accounts for this apparent disparity between the American and foreign orientation, between the seemingly continued vitality of tort liability in the one case, and its notable retreat in the other? Is America less animated by our era's concern for social welfare? Surely not, as the marked trend to strict tort liability itself demonstrates. Rather, the means chosen to promote this objective differ from those employed in most other countries. In them the principal building block is a national social security system, including a national health scheme. By contrast, the American preference for pluralism and the free market has led to a proliferation of compensation structures whose aggregate of benefits and expenditures can hold its own in comparison with the welfare efforts elsewhere.

The federal social security system, it is true, makes only a modest contribution to accident compensation, free hospital care being provided only for the elderly (Medicare) and the poor (Medicaid), and disability benefits only for persons (under 65) who suffer total and lasting disablement.[100] None the less, over 85 per cent of the population are covered by private medical insurance, though not all of these for long-term chronic illness.[101] In addition to workers' compensation, a number of states require non-occupational disability insurance,[102] many have enacted no-fault automobile insurance, and federal legislation has experimented with several compensation plans for special injuries, notably black lung disease[103] and now vaccine injuries,[104] as well as making itself

[100] Social Security Act 42 USC § 23: 'Inability to engage in any substantial gainful activity by reason of any medically determinable physical or mental impairment which can be expected to result in death or can be expected to last for a continuous period of not less than twelve months.'

[101] *Source Book of Health Insurance Data 1984–1985* (Health Insurance Assoc. of America) ch. 1.

[102] Cal., NY, NJ, RI, and Haw. See 'Social Security Programs', *supra*, n. 66, at 37; Report of the Research Subcommittee of the Disability Insurance Committee, Health Insurance Association of America, *Compensation Systems Available to Disabled Persons in the U.S.* (1978).

[103] A special federal programme administered by the Department of Labor since 1972 under the Federal Coal Mine Health and Safety Act.

[104] *Supra*, n. 52.

responsible for excess insurance for nuclear accidents.[105] Other special compensation plans have been proposed for a great variety of accidents, from medical[106] and toxic-chemical injuries[107] to extension of workers' compensation[108] or temporary disability insurance for twenty-four hour coverage.[109] Also notable is the O'Connell model of elective no-fault in exchange for tort benefits.[110] But these, let alone general accident compensation plans,[111] have remained utopian.[112] Characteristic of the existing mosaic of compensation sources is its decentralization and variety of funding. Private insurance plays a prominent role and contributions come from the insured, employers as well as taxpayers. In addition, voluntary insurance of various kinds provides an option for the risk-averse and those needing excess coverage.

Notwithstanding its fragmentation, there are advantages in

[105] Price–Anderson Act, 42 USC § 2210. Current proposals would raise industry liability from the current $660m. to $6.5bn..

[106] See Havighurst, 'Medical Adversity Insurance: Has Its Time Come?' 1975 *Duke L. J.* 1233; Tancredi, 'Designing a No-Fault Alternative', 49(2) *Law and Contemp. Prob.* 5 (1986); R. Henderson, 'The Boundary Problems of Enterprise Liability', 41 *Md. L. Rev.* 659 (1982).

[107] Soble, 'A Proposal for Administrative Compensation of Victims of Toxic Substance Pollution: A Model Act', 14 *Harv. J. Legis.* 683 (1977).

[108] R. Henderson, 'Should Workmen's Compensation be Extended to Nonoccupational Injuries?' 48 *Tex. L. Rev.* 117 (1969).

[109] Sugarman, 'Doing Away With Tort Law', 73 *Calif. L. Rev.* 555 (1985).

[110] O'Connell, *Ending Insult to Injury: No-Fault Insurance for Products and Services* (1975); id., 'An Elective No-Fault Liability Statute', 1975 *Ins. L. J.* 261; id., 'Giving Motorists a Choice between Fault and No-Fault Insurance', 72 *Va. L. Rev.* 61 (1986). The author is a co-sponsor of the Keeton–O'Connell automobile plan (*infra*, ch. 5, text near n. 91). In later years he has advocated elective plans to induce plaintiffs to trade-off the collateral source rule for the defence of contributory negligence (*infra*, ch. 6, n. 96) and damages for pain and suffering against their attorney's fees (*infra*, ch. 6, n. 183).

[111] It would be idle to expect support for such a radical solution from the American Bar Association (ABA). In 1987 it voted at last to support some modest reform, such as restrictions on awarding punitive damages and a limit on 'joint and several liability': *Nat. L. J.* 16 Feb., 2 Mar., 1987. A proposal to limit 'pain and suffering' awards was defeated. See also the conciliatory *Report to the ABA. Towards a Jurisprudence of Injury: The Continuing Creation of a System of Substantive Justice in American Tort Law* (1984).

[112] e.g. Bernzweig, *By Accident Not Design* (1980); Franklin, 'Replacing the Negligence Lottery: Compensation and Selective Reimbursement', 53 *Va. L. Rev.* 774 (1967).

this arrangement. Many of the insurances are first-party and adjustable to the individual's needs unlike the flat benefits under most centralized social security systems. Likewise, premiums are fine-tuned to reflect exposure to risk instead of being flat-rated and are in this and other respects more sensitive to the problems of 'moral hazard'. This contrast is particularly noticeable in the different approaches to motor-traffic accidents: while the traditional European method has followed the original workers' compensation model of third-party insurance or a central compensation fund, the unique American solution has been simply to convert voluntary into compulsory first-party insurance. This makes it possible to allow for a variety of deductibles and to counteract the tendency of liability insurance (and tort law) to subsidize the better-off. One of the ironies of the recent 'tort explosion' is a loss of this advantage as a result of shifting the balance toward third-party insurance. Indeed, one of the widely overlooked causes of the spectacular rise in insurance premiums has been the consequential need for redefinition and re-evaluation of these risks.

Thus, as elsewhere, tort liability plays only a complementary role to social and private insurance, one whose need to fill the gap is widely overestimated. Still, its proper function in the overall scheme stands in need of clarification. The so-called collateral source rule has the effect of generally permitting the accident victim to pocket both insurance benefits and unreduced tort damages, even when the insurance is mandatory and derived from public funds such as federal social security. This result could be condemned as a gross misallocation of resources but for the argument that successful tort plaintiffs have to earmark a substantial portion of their award for their attorney.[113] Inevitably, however, the function of tort law as a gap-filler is obscured, if not contradicted.

The overall accident compensation landscape in America is therefore highly fragmented and lacking any systematic plan. The benefits of tort liability are not husbanded for those who are un- or under-compensated from other sources but are distributed randomly without regard to the needs or deserts

[113] *Infra*, ch. 6, at n. 89.

of victims. Its grossly inflated transaction costs add to the remarkable inefficiency of the tort system. Only the affluence of American society has deflected serious misgivings about this wasteful arrangement and about the need to develop a more purposeful compensation policy such as that which other countries, with more limited resources, have been driven to pursue.

This brief overview underlines the importance of the broader context in which tort liability operates in the United States. What distinguishes the American tort system from its counterparts in other countries, including those of the common law, is not so much the substance of its 'black-letter' rules as its unique institutional arrangements. This is not to deny that social values or legal philosophies have left their mark on evolving doctrine and that differences in legal tradition can affect the direction and timing of legal change. But the institutional features of the system, themselves reflecting significant social values, account to a far greater extent for the peculiar distinctiveness of American tort law in operation. The following chapters will explore the more important of these: the relationship between courts and legislature as agents of reform (Chapters 2 and 3), the jury (Chapter 4), the lawyers and insurers (Chapter 5), the contingent fee (Chapter 6), and procedural efforts to cope with mass litigation (Chapter 7).

2

JUDICIAL ACTIVISM

1. THE JUDGES

Social change since the last world war, driven by heightened expectations of social welfare, affluence, and technological progress, has been putting the legal system under hitherto unparalleled pressure to accommodate itself to this changing vision. Tort law is peculiarly exposed to these demands in its role as the traditional source of accident compensation.

Legal systems in all Western countries have responded to this challenge to a varying extent and at varying speed. In the United States, more than elsewhere, it has fallen almost exclusively to the courts, unaided by legislation, to determine the direction and pace of legal adjustment. Whether willingly or not, an activist role was thus thrust upon American courts, especially courts of last resort. Although, for reasons to be presently explained, legislatures by and large yielded this high ground by default, the judicial role as reformer none the less inevitably trenches on the delicate division of functions between judge and legislator.

The wider discretion and creativity claimed by American, in comparison with British, courts is of course generally recognized and not peculiar to tort litigation. Notable institutional differences spring to mind: the sheer size and complexity of the judicial system and of the legal profession preclude the development of those shared professional values nurtured by the highly centralized and hierarchic British system. American judges, appointed primarily for their political affiliations and increasingly drawn from a wide spectrum of society and experience, lack both the intense professional training and the common ethos of the closely-knit English bar, from among the most outstanding of whom judges are selected. Accordingly, their loyalty is less to the Law as a closed system than to their personal ideology, to which they owe their appointment. Law, for them, is not an embodiment of 'neutral principles' of lasting truth, but an instrument of government to achieve goals for today and tomorrow. Their outlook remains essentially political, their concept of law 'instrumentalist'. Appellate judges 'vote' rather than adjudicate, vying to 'win' and if on the losing side hope that their dissent, often cast in strongly adversarial tones, will prevail another day.[1] This attitude is as endemic among appointed judges with long tenure as it is among those subject to short-term popular elections. 'Judicial independence' does not everywhere carry the message that the price for freedom from politics should be judicial restraint and abnegation of political law-making power.

Certainty of the law occupies a relatively lowly place in the axiology of American jurisprudence. British courts mostly content themselves with interstitial gap-filling and disclaim any power to overrule precedents because they consider them to have been wrongly decided.[2] Their province is the development of 'principle', not 'policy' (though this distinction is not without its own problems).[3] By contrast, the optimistic

[1] Even if, as Griffith among others contends, English judges have a conservative bias, they at least believe that they are pursuing 'neutral principles'. (Notably, Griffith does not complain of their record in tort cases.) Griffith, *The Politics of the Judiciary* (3rd edn. 1985). See also Atiyah, 'Lawyers and Rules: Some Anglo-American Comparisons', 37 *Sw. L. J.* 545 (1983).

[2] This is not the place for going into this much debated topic beyond giving a few references: Paterson, *The Law Lords* (1982); Stevens, *Law and Politics* (1979); (Lord) Reid, 'The Judge as Lawmaker', 12 *J. SPTL* 22 (1972); (Lord) Devlin, 'Judges and Lawmakers', 39 *Mod. L. Rev.* 1 (1976). English concern with *stare decisis* has spawned its own considerable literature, with no American counterpart: Stone, *Precedent and Law: Dynamics of Common Law Growth* (1985); Cross, *Precedent in English Law* (3rd edn. 1977).

[3] Highlighted by the confrontation between Lords Scarman and Edmund-Davies in *McLoughlin* v. *O'Brian* [1983] 1 AC 410. It also shows up in the ambiguities of Ronald Dworkin's 'rights thesis' which may have influenced Lord Scarman's repudiation of 'policy' as much as did his distinguished association with the Law Reform Commission. Dworkin's very objective of limiting judicial discretion in choosing legal goals (*Taking Rights Seriously*, 1979), lends itself easily to subversion by promoting broad social goals, including the redistribution of wealth and entitlements, under the pretence of following the principle of foreseeability. Indeed, by accepting that 'rights' can make collective goals relevant, he seems to give tacit assent

American belief in the virtue of progress entails a cultural commitment to change, which endows judges no less than other decision-makers with a mission to improve the existing order. *Stare decisis* reflects a static, traditionalist society; legal innovation a society 'on the move'.

Tort law is probably more susceptible than most other branches of law to this appeal, partly because it is more influenced by changing public perceptions, partly because it engages less reliance on the status quo. Moreover, the very structure of tort law, expressed in open-ended 'principles' rather than 'rules' (reasonableness, foreseeability), invites an infusion of 'policy' and contemporary social values from the decision-makers: judges and juries. Indeed, American experience clearly reveals that blind pursuit of the 'principle' of foreseeability can become a screen for activist judges deliberately engaging in the role of broad-ranging social reform.[4] Conversely, appeal to 'policy' can serve no less as a brake than as an accelerator of legal change.

Tort law in the United States is primarily state law.[5] State supreme courts have the last word on the substantive law of

to the argument that economic policies like cost-benefit analysis and loss distribution may become legitimate in defining the rights of negligence victims.

[4] For example in California where foreseeability became the 'key component' under the *Bird* court: see *J'Aire Corp.* v. *Gregory*, 24 Cal. 3d 799, 806, 598 P. 2d 60 (1979). Cf. this comment on a similar English tendency: 'The *Anns* [i.e. foreseeability] approach is attractive to a wide range of judges, producing a strange alliance of conceptualists concerned with doctrinal purity and form, social engineering activists who see the tort of negligence as a valuable welfare tool for income maintenance and wealth distribution, utilitarian pragmatists concerned with efficient allocation of resources and risks on a case-by-case basis, and judges under pressure concerned primarily with reaching a decision on the facts which is "appeal-proof" ' (Smillie, 'Principle, Policy and Negligence', 11 *NZU L Rev.* 111, 148 (1985)). See also Bell, *Policy Arguments in Judicial Decisions* (1983).

[5] Prior to *Erie R. R. Co.* v. *Tomkins*, 304 US 64 (1938) federal courts tended to develop a federal 'common law' in diversity cases; hence some of the notable pronouncements on tort doctrine by Justice Holmes. Since 1938 federal courts must apply state law in diversity cases, i.c. where one litigant claims a federal *forum* against a citizen of another state.

Federal tort law is confined to a few areas, e.g. torts on the high seas, on navigable waters, and under the Federal Employers' Liability Act (FELA) dealing with accident claims by railway employees.

their jurisdiction except for the limited restraints of the federal Constitution.[6] In turn, state judges reflect the political culture of their locality. Conservative, mainly rural states, select judges committed to strict construction and *stare decisis*; 'liberal', primarily urban, states tend to elevate more jurists imbued with a mission for social change. The former see themselves primarily as adjudicators, the latter as lawmakers.[7] Even within any one appellate court, judges are often of diverse background and orientation, reflecting a divided electorate and political swings in the selection process. This in turn accounts for the factionalism and high prevalence of dissents in some courts.[8] Studies of voting records of individual state Supreme Court judges confirm the high correlation between their political affiliation and decisions. Liberal Democrats in tort cases generally support claimants (in criminal cases defendants); Republicans vote the opposite way.[9] Since courts of last resort control their own docket, the majority judges are able by pointed selection of appeals to control also the speed and direction of legal change. Significantly, the percentage of tort appeals in most jurisdictions during the period under review has overshadowed all but criminal and public law cases.[10] Few appeals are routine.

The cultural and ideological diversity among states has been softened in its long-term effect on legal change by the pervasive influence of accumulating precedents in sister jurisdictions. While state Supreme Courts are final arbiters of local law, they cannot escape the dynamic of evolving consensus elsewhere. Hence the importance attached to opinion-leaders such as the Supreme Courts in the major 'forward-looking'

[6] See *infra*, ch. 3.

[7] Glick, *Supreme Courts in State Politics*, ch. 2 (1971) for comparative attitudes between NJ, Mass., Pa., and La. judges.

[8] Ibid., ch. 5 ('decision-making and group interaction').

[9] See Nagel, 'Political Party Affiliation and Judges' Decisions', 55 *Am. Pol. Sc. Rev.* 843 (1961); Glick and Vines, *State Court Systems* 50-1, 82-5 (1971), reporting that 75.5% of judges held at least one nonjudicial political office, on average 2.3 offices (pp. 48-9). The doyen of judicial 'role' literature is G. Schubert whose (quantitative) studies have dealt mainly with the US Supreme Court: Schubert, *The Judicial Mind* (1965); *The Judicial Mind Revisited* (1974).

[10] See e.g. Glick and Vines, *supra* 98, on statistics for 1965-6.

jurisdictions, like California, Florida, Michigan, New Jersey. More conservative courts come under progressive pressure to fall into line, and are able to justify eventual break with their own precedents by pleading the irresistible pull toward national conformity.[11]

Judicial activism, especially when pursued in brash and arrogant fashion, exacts a high price in divisiveness, politicization, and disrespect for governmental institutions. Tort and criminal law not only claim the vast bulk of all litigation but also share a high level of public attention due to widespread disagreement about social purposes and means. This contentiousness in the case of torts is fanned by special interest groups: the plaintiffs' bar and consumer organizations on the one hand and a combination of target defendants (business and professions) and the insurance industry on the other. This contentiousness tends to spill over into politics, in turn colouring the selection of judges and the judicial process itself. In attempting to capture that process in these pages, a first step is to look at the pace and methods of judicial reform of tort law during the past three decades.

The American law of torts had assumed an anachronistic look by the middle 1950s. By comparison, long overdue legislative reforms in England, followed in most instances in other Commonwealth jurisdictions,[12] had in the period around the Second World War adapted the law to the changed demands of the mid-century. That the long cycle of stagnation there[13] was at last drawing to a close was first evidenced in the torts field by the legislation in 1934 allowing survival of actions[14] and that in 1935 introducing contribution between tort-

[11] See Baum and Canon, 'State Supreme Courts as Activists: New Doctrines in the Law of Torts', *State Supreme Courts: Policy-Makers in the Federal System*, ch. 4 (1982) which ranks the 50 courts by their innovative record. For a rare, sustained criticism of the California Supreme Court's activism by another judge see Fernandez, 'Custom and the Common Law: Judicial Restraint and Lawmaking by Courts', *11 Southw. U. L. Rev.* 1237 (1979).

[12] In some instances, anticipated in the Commonwealth, e.g. apportionment of loss in Ontario and some other Canadian provinces.

[13] See Abel-Smith and Stevens, *Lawyers and the Courts*, pt. 2 (1875-1939: The Era of Stagnation) (1967).

[14] Law Reform (Miscellaneous Provisions) Act 1934.

feasors.[15] There followed shortly after the war legislation, first to replace the defence of contributory negligence by apportionment of loss in 1945[16], and second to abolish Crown immunity in 1947.[17] Occupiers' liability was brought under the embracing standard of the common duty of care in 1957,[18] the immunity of highway authorities fell in 1961,[19] followed by the final remnants of marital immunity in 1962.[20] These were all statutory reforms, while the courts during that period contented themselves in the main with modest refinements of established principles.[21] Not until the appointment of the Pearson Commission in 1972 was the need felt to take a more critical look at the general panorama of accident compensation, especially the relationship between tort law and social security which had burgeoned under the impetus of the Beveridge Report of 1943.

In the United States, as already mentioned, the task of similarly modernizing the law of torts has fallen to the courts. Starting in the late 1950s, one after another of the anachronistic features of tort law were dismantled, first the immunities—family, charitable, and sovereign—later the introduction of comparative fault both as between plaintiff and defendant (contributory negligence) and between defendants (contribution) and, most dramatic of all, strict liability for defective products. These doctrinal changes laid the groundwork for the remarkable transformation of the American law of torts from a relatively modest to an aggressively expansive role in accident compensation. In general, legislation played, at best, only a marginal, reactive role in this process.

In tracing this development, we will first seek an explanation for the prevailing legislative inertia in the structural condition of the American legislative process. I will then illustrate the peculiar interaction between court and legislature in their role

[15] Law Reform (Married Women and Tortfeasors) Act 1935.
[16] Law Reform (Contributory Negligence) Act 1945.
[17] Crown Proceedings Act 1947.
[18] Occupiers' Liability Act 1957.
[19] Highways (Miscellaneous Provisions) Act 1961.
[20] Law Reform (Husband and Wife) Act 1962.
[21] See e.g. Stevens, *supra*, pt. 4 (1956–76). The author lauds an 'incremental development of new doctrines' (589) in the mid-1970s, but that was small beer compared with the American brew.

as lawmakers by focusing on different episodes involving changes both of the common law and of statute law.

2. THE LEGISLATIVE PROCESS

Some twenty-five years ago, Professor Cornelius Peck set out to explore the reasons for the creative role assumed by American courts in reforming the law of torts and concluded that, contrary to widespread myths, the comparative abilities for this task favoured courts over legislatures on many, perhaps most, of the issues that had called for resolution during this period.[22] While some developments, such as the lobbying influence of the plaintiffs' bar, have changed in the meantime, the major thrust of his analysis remains today as true as it is elucidating.

2.1   Inertia

Legislative indifference to any programme of tort reform is perhaps the strongest argument to silence protests about judicial usurpation. Legislators are absorbed by the problems of passing a budget, responding to daily pressures from constituents, clients, and lobbyists, and readying themselves for the next election every two years. Even in large states, such as New York and California, where legislation is no longer a part-time occupation and sessions take place every year, no longer in alternate years, legislators in general lack both the staff resources and the incentives to devote themselves to any sustained task like principled law reform. Some states now have law reform commissions, but these are dependent on specific legislative directives, are part-time and generally ineffective. The Attorney General's department oversees criminal law enforcement and advises the Government, but has no more inclination than the executive (the Governor) to monitor the operation of substantive civil law, let alone launch any ambitious programme of law reform. The Uniform Law Commissioners, composed of state delegates, have promulgated

[22] Peck, 'The Role of the Courts and Legislatures in the Reform of Tort Law', 48 *Minn. L. Rev.* 265 (1963).

a not inconsiderable volume of model laws, whose rate of adoption, however, is dispiriting and reflects the pervasive lack of disinterested concern about the condition of private law.[23]

On the very rare occasions when a more ambitious review is launched, it is as likely to be gutted by competing interests and lapse into coma. One such example was the Joint Committee of the California Legislature on Tort Reform, set up in 1977 to quieten a swell of adverse publicity and protests about the increasing burden of tort liability in the wake of the medical insurance crisis, alarming awards of compensatory and punitive damages in products cases, and so on.[24] Studies from experts were commissioned, but the enthusiasm of the legislative command structure quickly evaporated, giving sustenance to the suspicion that the whole effort was mere window-dressing from the start.[25]

Such legislation in the torts field as intermittently passes into the statute book is usually a response to a momentary crisis, unrelated to any wider principle and thereby becoming hostage to constitutional attack on grounds of discrimination.[26] Examples are statutes imposing liability in limited amounts on parents for wilful destruction of property by minor children[27] or an occasional reversal of a court decision that has touched a sensitive nerve in constituents, like the decision imposing liability on a social host for supplying excessive alcohol to a guest who later causes a driving accident on the way home.[28] All this is of course a far cry from the systematic watch of law reform agencies in many Continental and, more recently, Commonwealth countries. In the former, Ministries

[23] The most important model acts in the torts field are the Uniform Contribution Among Tortfeasors Act (1939 and 1955), the Comparative Fault Act (1979), and the Periodic Payment of Judgments Act (1980). Of these the first has had scattered adoption, the last two none.

[24] See e.g. Report of the California Citizens Commission on Tort Reform, *Righting the Liability Balance* (1977).

[25] See *Series 1978 Staff Report* (1979) comprising studies and recommendations. A few were adopted by the court, none by the legislature to whom they were addressed.

[26] *Infra*, ch. 3.

[27] See Prosser and Keeton, *Law of Torts* 913 (5th edn. 1984).

[28] *Coulter* v. *Superior Court*, 21 Cal. 3d 144, 577 P. 2d 669 (1978) was promptly repealed by Bus. & Prof. Code § 25602.

of Justice are charged with continuing supervision over the operation of the civil as well as of the criminal law, and are dedicated guardians of the integrity of their civil code. Indeed, it was Judge Cardozo who, as long ago as 1921, strongly recommended this model for the United States.[29] In England (and Scotland), the permanently staffed Law Commission has accomplished a laudable record of law reform measures since its establishment in 1965,[30] successively emulated in Canada, Australia, and elsewhere in the Commonwealth.[31]

2.2   Lobbies

Widespread belief has it that legislatures command better access to information and are thus in a superior position, compared with courts, to assess the needs and merits of proposed reforms. This assumption does not hold true, at least as a generalization. On many issues, particularly of lawyers' law, judges' expertise could not be rivalled by non-professionals, even after making allowance for the presence of lawyers among legislators, especially on judiciary committees.

The committee system, admittedly, provides an opportunity for interested parties to present their views and hope for publicity that might give additional weight to their arguments. But most observers of the process are sceptical whether this routine is more than a charade to satisfy democratic pretensions. Legislators' minds are far more likely to be swayed by the private intercession of lobbyists than by the highly orchestrated public testimony of the spokesmen for opposing views.

Reform of tort law in the American context assumes a highly adversarial posture. The reason for this is that any change in the prevailing equilibrium of legal rights is bound to prejudice either plaintiffs or defendants. Unlike in other countries, both these causes are nowadays powerfully represented by opposing organizations of lawyers: on the one

[29] Cardozo, 'A Ministry of Justice', 35 *Harv. L. Rev.* 113 (1921).

[30] Law Commissions Act 1965. See Smith and Bailey, *The Modern English Legal System* 18–20 (1984).

[31] See *Reform*, 'a regular bulletin of law reform news, views and information, published by the Australian Law Reform Commission'.

hand, the trial lawyers; on the other, the defence bar and its clients. As will be explained in a later chapter in greater detail,[32] the former group has a vital stake in the outcome since it is apt to affect directly their contingent fees. The American Trial Lawyers Association (ATLA) and its local affiliates consequently maintain a powerful legislative watch and its influence on the legislative process has become impressive.

ATLA is usually opposed by a number of organizations representing potential tort defendants, like the American Manufacturers Association on issues of products liability, and of course liability insurers. This defence group is not as cohesive as the plaintiffs', usually forming temporary alliances on specific issues, as with the medical association and pharmaceutical manufacturers on the question of vaccine liability or with the various media organizations on libel.

The effect of this highly structured confrontation is to substantially paralyse legislative reform in the torts area. This has not always been so. Historically, state legislatures were in general dominated by conservative majorities and interests. Prominent among the reasons was the disproportionate electoral strength of the rural vote before the constitutional requirement of one man, one vote in 1962,[33] the enfranchisement of minorities,[34] and the dramatic rise of ATLA in the post-war period. It is from those early years that defence-orientated legislation, like the notorious guest statutes, stemmed.[35] Nowadays, however, the status quo has become difficult to shift either way. This may work against either side, depending on who is the proponent of change. Thus efforts by a section of the insurance industry to pass a strong automobile no-fault bill containing restrictions on tort recovery have been consistently thwarted by trial attorneys in California and many other states.[36] By the same token, however, almost all bills seeking to reverse or restrict pro-plaintiff decisions by the erstwhile radical majority of the California Supreme Court

[32] *Infra*, ch. 5.
[33] *Baker* v. *Carr*, 369 US 186 (1962).
[34] As the result of the Civil Rights Act of 1964.
[35] *Infra*, ch. 3 at n. 12.
[36] *Infra*, ch. 5 at n. 75.

have also been doomed.[37] In the event, the consoling thought that the legislature might always have the last word in controlling an unruly judiciary has proved largely fanciful.

This leads to another paradox. One might have argued in the abstract that legislative paralysis provided a sufficient justification for judicial initiatives.[38] Yet this very paralysis tends to weaken the conventional self-restraint on the part of judges bent on promoting radical social reforms without fear of reversal or retribution. This tendency entails two serious detrimental consequences. One is that it further undermines the judicial claim to legitimacy. Democratic societies supposedly view with suspicion anti-majoritarian pretensions to power. Even if the constitutional separation of powers in the United States assures independent status to the judiciary, co-ordinate with the two other branches of government, the conventional assumption has been that courts do not usurp the lawmaking function of the legislature. While a clear line cannot be drawn between making and applying law and that line is in any event conditioned by a subtly changing balance between court and legislature, any radical reforms of broader social implications by the judiciary are bound to be widely perceived as illegitimate and unconstitutional.

This in turn politicizes the court and detracts not only from its claim to legitimacy but also alienates it from broad sections of the community. The California experience in recent years, in which justices of the Supreme Court were spectacularly defeated at the polls,[39] testifies to this unfortunate but not surprising public perception.

[37] Notable have been futile industry efforts to reverse *Sindell* v. *Abbott Laboratories*, 26 Cal. 3d 588, 607 P. 2d 924 (1980) (the market-share decision) and *American Motorcycle* v. *Superior Court*, 20 Cal. 3d 578, 578 P. 2d 899 (1978) (retaining the 'joint and several liability' rule). The first is discussed *infra*, ch. 7, at n. 92; the second *infra*, ch. 3, at n. 76. A bill to modify *Rowland* v. *Christian*, 69 Cal. 2d 108, 443 P. 2d 561 (1968), so as to disqualify felons from suing occupiers of land, was eventually enacted in 1985 (Civ. Code § 847).

[38] 'When such a stalemate exists and the legislature has, for whatever reason, failed to act to remedy a gap in the common law that results in injustice, it is the imperative duty of the court to repair that injustice': *Avis* v. *Ribar, infra*, n. 64, at 23, 431 NE 2d at 896.

[39] 5 Nov. 1986. These judges stood for re-election to a 12-year term. For earlier episodes see Stolz, *Judging the Judges* (1981).

3. JUDICIAL LAWMAKING

3.1 Prospective overruling

Linked to the old-fashioned idea that courts do not make law but merely 'discover' or apply it, is the corollary that, while statutes mostly operate only prospectively,[40] judicial decisions operate retroactively under the pretence that even an overruling of earlier precedent merely corrects an improper understanding of what the law has always been. Notwithstanding this fanciful Blackstonian gloss, the conventional retroactive effect of decisions has had a cautionary effect on judicial initiatives especially in areas where reliance on precedent controls human behaviour, as in the case of property dispositions and commercial practices. Human conduct may not be greatly influenced by specific rules of tort law, but defensive measures by potential defendants, like procuring liability insurance, are. In order to mitigate this concern, and thereby to gain greater confidence for reform, American courts have invented the technique of prospective overruling.[41]

This technique owes its widescale acceptance principally to its sponsor, Judge Cardozo, who advocated it repeatedly in extrajudicial addresses before being vouchsafed the personal triumph of adopting it on behalf of the US Supreme Court. As he pointed out in the 1921 Storrs Lectures:

In the vast majority of cases the retrospective effect of judgemade law is felt either to involve no hardship or only such hardship as is inevitable where no rule has been declared. I think it is significant that when the hardship is felt to be too great or unnecessary, retrospective operation is withheld. Take the cases where a court of final appeal has declared a statute void, and afterwards, reversing

---

[40] But occasionally enacted retroactive operation is not unconstitutional: see *In re Consolidated U.S. Atmospheric Testing Litigation*, 616 F. Supp. 759 (DC Cal. 1985). Only attempts to alter the rule of decision in pending cases violates Art. III (separation of powers): *US* v. *Sioux Nation of Indians* 448 US 371, 404 (1980).

[41] See especially Schaefer, 'The Control of "Sunbursts": Techniques of Prospective Overruling', 42 *NYU L. Rev.* 631 (1967).

itself, declares the statute valid. Intervening transactions have been governed by the first decision . . . Most courts in a spirit of realism have held that the operation of the statute has been suspended in the interval . . . Why draw the line here? Where the line of decision will someday be located, I will make no attempt to say. I feel assured, however, that its location, wherever it shall be, will be governed, not by metaphysical conceptions of the nature of judge-made law, nor by the fetish of some implacable tenet, such as that of the division of governmental powers, but by considerations of convenience, of utility, and of the deepest sentiments of justice.[42]

Prospective overruling, although occasionally employed before, was endorsed in 1932 by the US Supreme Court in *Great Northern Railway* v. *Sunburst Oil & Refining Co.*,[43] from which it took its name, commonly used, of 'sunbursting'. It has not been confined to decisions on the validity or construction of statutes, but has been frequently invoked in the torts field, especially in abrogating governmental,[44] charitable,[45] and interfamily immunities.[46] The reason for resorting to this strategy was, of course, to give potential defendants a chance to take protective steps like obtaining liability insurance and raising their prices so as to pass on the cost to taxpayers, customers, and clients.

Prospective overruling assumed an additional nuance when the Minnesota Court postponed its abrogation of sovereign immunity until the end of the next session of the legislature.[47]

[42] Published as *The Nature of the Judicial Process* 146–9 (1921). By contrast, the persisting unwillingness of the House of Lords to make this 'yet further innovation' was cited as a reason against making even a modest extension of vicarious liability (like the American 'family car' doctrine): Lord Wilberforce in *Launchbury* v. *Morgans* [1973] AC 127, 137. This speech by a centrist judge, explicitly addressing the general issue of innovation (without even the necessity of overruling), provides a poignant contrast to the prevailing American methodology.

[43] 287 US 358.

[44] e.g. *Molitor* v. *Kaneland Community Unit Dist. No. 302*, 18 Ill. 2d 11, 163 NE 2d 89 (1959). See the extended analysis in Mishkin and Morris, *On Law in Courts* 272–317 (1965).

[45] e.g. *Parker* v. *Port Huron Hosp.*, 361 Mich. 1, 105 NW 2d 1 (1960).

[46] e.g. *Goller* v. *White* 20 Wis. 2d 402, 122 NW 2d 193 (1963).

[47] *Spanel* v. *Mounds View School Dist.*, 264 Minn. 279, 118 NW 2d 795 (1962). Even more circumspect was the Mass. Court in *Whitney* v. *City of Worcester*, 373 Mass. 208, 366 NE 2d 1210 (1977), merely declaring its intention to abrogate the immunity in the first case to come up after the

It thereby purposely expressed its deference to the legislature as the ultimate lawmaking authority and, while suggesting its own view as to the need for reform, clearly invited the legislature to review its decision. The Minnesota legislature did just that and enacted a statute which, *inter alia*, retained the immunity of school districts for another four years and qualified the new liability of municipalities in several ways.[48] This example illustrates the greater sensitivity to fiscal implications affecting the taxpayer or strongly organized interests like local government bodies. By contrast, the Minnesota Court did not deem this caution necessary when overruling (prospectively) the immunity between parent and child a few years later.[49] The same concern is reflected by the legislative response in California to the Court's abrogation of governmental immunity.[50] In contrast to the usually prevailing inertia, the legislature in that instance promptly suspended the decision for two years, thus gaining time to initiate a detailed study of the problem and eventually pass a complex and detailed government liability statute.[51]

If prospective overruling were to deny the immediate litigant the fruits of victory, it would not only reduce the court's decision to mere dictum—and courts have not always adhered to their predictions[52]—but it would seem both unfair[53] and to take away all incentive for litigants to press for a change of the law. To meet these objections, courts frequently apply the new ruling to the successful party as a reward for stimulating judicial reform but otherwise postpone its application to events occurring after the date of the decision, for

conclusion of the next legislative session, but then making it retroactive to 1973 when the Court first signalled its intention to reconsider the doctrine.

[48] Minn. Stat. Ann. §§ 466.01–6.12.

[49] *Balts* v. *Balts*, 273 Minn. 419, 142 NW 2d 66 (1966).

[50] *Muskopf* v. *Corning Hospital Dist.*, 55 Cal. 2d 211, 359 P. 2d 457 (1961).

[51] See Van Alstyne, 'Government Tort Liability: Judicial Lawmaking in a Statutory Milieu', 15 *Stan. L. Rev.* 163 (1963); id., 'Governmental Tort Liability: A Public Policy Prospectus', 10 *UCLA L. Rev.* 463 (1963).

[52] e.g. *Scheele* v. *City of Anchorage*, 385 P. 2d 582 (Alaska 1963) (revg. itself on prospective overruling of sovereign immunity).

[53] But the US Supreme Court in the *Sunburst* case (*supra*, n. 43) expressly held that the practice did not violate due process.

example to accidents or trials thereafter.[54] This compromise 'may be a bit intellectually untidy, but it may offer the best alternative to judicial paralysis on the one hand, and wholesale frustration of legitimate expectations on the other'.[55]

3.2   Comparative fault

Notwithstanding a judicial inclination to dispose of the problem of overruling with a broad assertion that a judge-made rule can always be undone by judges, the problem calls for, and often receives, more differentiated treatment. Some of the factors relevant to the determination whether the court or the legislature is better qualified to accommodate the need for change have already been mentioned: familiarity with the specific problem, the influence of pressure groups, fiscal implications. Others, like the comparative capacity to set up a complex alternative structure, the public contentiousness of the issue, and the preclusive role of prior legislative concern with the problem were additionally involved in the question whether to replace the defence of contributory negligence by comparative fault (apportionment of loss). Because the arguments either way were peculiarly balanced and judicial responses accordingly mixed, this debate justifies more extensive attention.

By mid-century judicial and most professional support for the defence of contributory negligence had eroded in the United States as in other countries of the common law. Juries frequently ignored it, the courts had strained to engraft exceptions like the 'last clear chance' rule, and some statutes had abrogated it either generally or, more commonly, for specific situations.[56] In its origin and heyday, the defence may

---

[54] e.g. *Molitor* (*supra*); and *Balts* (*supra*) dealing with immunities; *Li* v. *Yellow Cab* (*infra*, n. 66) and *Alvis* v. *Ribar* (*infra*, n. 64) dealing with comparative fault. Think what this says about the courts' value of judicial activism! But might not a lesser incentive than full recovery have been sufficient? Ask Mishkin and Morris, *supra* at 310–14.

[55] Currie, 'Suitcase Divorce in the Conflict of Laws', 34 *U. Chi. L. Rev.* 26, 61 (1966), cited by Schaefer (*supra*).

[56] e.g. Prosser and Keeton, *supra*, ch. 12; Harper, James, and Gray, *Torts*, ch. 22 (2nd edn. 1986). Statutes of general application obtained in Ark., Miss., Me., Nebr., S.Dak., and Wis. The prominent example of statutes of

well have served a calculated social and economic function by reducing the cost of compensation in the interest of building a stronger economy.[57] Yet not only had industry long ceased to be regarded as in need of such a subsidy, but there was now a broad consensus that compensation of accident victims was a desirable social goal that could be accomplished by loss-spreading through the conduit of liability insurance and ultimately passing the cost to the consumer public. From two of the three important accident areas—work injuries[58] and defective products[59]—the plaintiff's contributory fault had been virtually expelled, so that the defence in any event played only a lingering role in road and other residuary accidents. An additional reminder of its incongruity was the well-publicized fact that, as the civil law countries of the world had long rejected their Roman law heritage, the all-or-nothing rule, followed recently by most Commonwealth legislation,[60] America had become the last stronghold of this now anomalous rule.

Its sole defenders—insurers and their clients—rested on the prognosis that a change to comparative negligence (apportionment) would lead to a substantial increase in insurance rates by drastically changing the tactical balance in settlement negotiation.[61] But at a critical point a conjunction of events suddenly came to favour the cause of reform.

Coinciding with the increased legislative muscle of the trial

special application is the Federal Employers Liability Act (FELA) regulating railways' liability to their own employees (1909), and state statutes modelled thereon.

[57] See Malone, 'The Formative Era of Contributory Negligence', *Essays on Torts*, ch. 4 (1986), emphasizing its role in railway accidents.

[58] By workers' compensation. A few industries remained under the common law, most prominently railways (*supra*, n. 56), agricultural employment in some states, but these had usually introduced comparative fault.

[59] Contributory negligence, short of assumption of risk, is not generally recognized as a defence to strict products liability: Prosser and Keeton, *supra*, § 102; Harper, James, and Gray, *supra*,§ 22.7.

[60] See Fleming, *Law of Torts*, ch. 12 (7th edn. 1987).

[61] The record, comparing costs between common law and comparative negligence jurisdictions, did not fully bear out this prognosis, pointing to a higher proportion of plaintiffs' verdicts, though not to larger awards. See Rosenberg, 'Comparative Negligence in Arkansas: A "Before and After" Survey', 13 *Ark. L. Rev.* 89 (1959).

attorneys (ATLA), defence spokesmen abruptly proclaimed their willingness to abandon their traditional opposition to reform: in defence of the common law tort system, they were now prepared to admit the plausibility of some improvements, including possibly a qualified rule of comparative negligence, in order to avert the greater evil of automobile no-fault.[62]

The assault on contributory negligence was mounted both in legislatures and in the courts. The possibility of judicial intervention began to be taken more seriously in the early 1960s when appeals to the legislature still looked forlorn.[63] But the first attempt failed disappointingly, when in 1968 the Illinois Supreme Court first remitted the issue to the Court of Appeals for consideration but, after receiving an affirmative reply on the merits, lost courage and, by a five to two majority, reaffirmed the old rule on the ground that the change had to await legislative action.[64] The Florida Court earned the distinction of becoming the first state, in 1973, to impose comparative negligence by judicial fiat.[65] Two years later the California Court lent its prestige to the same initiative.[66] In the meantime, twenty-one statutory adoptions had since 1969 joined the small group of six pioneers who, more than half a century earlier, had enacted comparative negligence statutes of general application.[67] By 1987 some forty states have adopted the reform; more than ten by judicial decision, two of them as recently as 1984.[68]

[62] See further *infra*, ch. 5, at n. 75. In this posture, they were joined by the trial bar (plaintiffs' attorneys): see Krause, 'No-Fault's Alternative: The Case for Comparative Negligence and Compulsory Arbitration', 44 *N. Y. St. B. J.* 535 (1972).

[63] Thus in 1962 Professor Rob. Keeton published a model judicial opinion for introducing comparative negligence, as a lure for courts endowed with more boldness than literary confidence. Keeton, 'Creative Continuity in the Law of Torts', 75 *Harv. L. Rev.* 463, 508–9 (1962). Even Prosser, *Law of Torts* 445 (3rd edn. 1964) did not rule out the possibility.

[64] *Maki* v. *Frelk*, 40 Ill. 2d 193, 239 NE 2d 445 (1968). That decision was ultimately reversed in *Alvis* v. *Ribar*, 85 Ill. 2d 1, 421 NE 2d 886 (1981).

[65] *Hoffman* v. *Jones*, 280 So. 2d 431 (Fla. 1973).

[66] *Li* v. *Yellow Cab Co.*, 13 Cal. 3d 804, 532 P. 2d 1226 (1975).

[67] See *supra*, n. 56.

[68] A list of States is found in *Alvis* v. *Ribar, supra*, at 12–14, 421 NE 2d at 891–2. The latest were *Hilen* v. *Hays*, 673 SW 2d 713 (Ky. 1984) and *Langley* v. *Boyter*, 325 SE 2d 550 (S. C. App. 1984). See generally Schwartz, *Comparative*

Were these courts justified in assuming the task of reform? To start with, there was no denying the centrality and entrenchment of the common law all-or-nothing rule. As Bohlen, father of the first *Restatement of Torts* once put it, 'this conception is part of the very atmosphere of English legal thought'.[69] It was not just a rule of marginal application, such as that dealing with emotional disturbance[70] or pre-natal injuries,[71] but one combining longevity with ubiquity. Thus, no useful analogy could be drawn to the judicial origin of comparative negligence in France where the eventual decision put to rest half a century's judicial vacillation,[72] or in the Soviet Union where the decision came soon after the promulgation of the first Soviet civil code and was remarkable only because that code gave no specific direction in the matter.[73]

That this 'core' aspect was once regarded as prohibitive by the California Court itself is demonstrated by its reaffirmation, not many years earlier, that a deceased's contributory negligence barred the wrongful death claim of survivors. The court had so held on the ground that the rule, even if difficult to square with the relevant statute (Code of Civil Procedure Section 377), was of such long standing that it had the acquiescence of the legislature and could not be touched by the court.[74] Now, however, this reasoning no longer carried

Negligence (1974); Woods, *The Negligence Case: Comparative Fault* (1978); 78 ALR 3d 339 (1977).

[69] Bohlen, 'Contributory Negligence', 21 *Harv. L. Rev.* 233, 253 (1908).

[70] Many jurisdictions have abandoned the 'impact' rule, some were in the process of dismantling also the requirement that the plaintiff must have feared for his own safety. See *Dillon* v. *Legg*, 68 Cal. 2d 728, 441 P. 2d 912 (1968). Breitel, J. in *Tobin* v. *Grossman*, 24 NY 2d 609, 249 NE 2d 419, 421 (1969) declined to take the second step, distingishing between 'only expansions' and 'significant creations of entirely new causes of action'. This was the less persuasive as he placed the decision giving wives a claim for loss of *consortium* into the first category. Others have sanctioned (or rejected) even filial claims without reference to such an elusive criterion. See generally Prosser and Keeton, *supra*, § 54, Harper, James, and Gray, *supra*, § 18.3.

[71] See Prosser and Keeton, *supra*, § 55; Harper, James and Gray, *supra*, § 18.3; 40 ALR 3d 1222 (1971).

[72] See Turk, 'Comparative Negligence on the March', 28 *Chi.-Kent L. Rev.* 189 (1950).

[73] See Rudden, 'Courts and Codes in England, France, and Soviet Russia', 48 *Tul. L. Rev.* 1010, 1023-4 (1974).

[74] *Buckley* v. *Chadwick*, 45 Cal. 2d 183, 288 P. 2d 12 (1955).

the day. Instead, the *Li* Court invoked the Civil Code itself as a mandate for changing the law. Its reasoning was a veritable *tour de force*. For after conceding that the only relevant Code section, Section 1714, announced the basic rule of negligence coupled with a complete defence of contributory negligence,[75] the court discerned nevertheless a pervasive legislative intent that the codification of 1872 not impede further development of these cardinal concepts. From the specific Code instruction to construe provisions codifying existing statutes and common law as continuations thereof,[76] the court deduced authority for an ongoing judicial evolution of these provisions beyond the confines of the statutory language itself.[77] Thus, in one fell swoop, it demolished both the argument that comparative fault was pre-empted by existing statute law and the argument that, in any event, the all-or-nothing rule was too interwoven in the very fabric of the law to be set aside without express legislative sanction.

A complementary factor was that the all-or-nothing rule had become haphazard in application, riddled by ill-defined exceptions (last clear chance, fault difference in kind not merely in degree) and the ameliorative judgment of juries. In the same way governmental immunity had become under-mined by the distinction between governmental and pro-prietary functions[78] and the charitable immunity of hospitals by the distinction between administrative and professional duties.[79] In short, by abrogating the old rule, the court was merely continuing a long process of erosion, giving the last push to a moribund institution. This appeal to rationality has become a characteristic technique of justifying overruling by American courts. By resolving systemic inconsistencies ('jagged doctrines'), the court can claim that the values underlying the standard of doctrinal stability and the principle of *stare decisis*—

[75] 'Everyone is responsible, not only for the result of his willful acts, but also for an injury occasioned to another by his want of ordinary care or skill in the management of his property or person, except so far as the latter has, willfully or by want of ordinary care, brought the injury upon himself.'

[76] See Civil Code § 4.

[77] Criticized by England, *Li* v. *Yellow Cab Co.*: 'A Belated and Inglorious Centennial of the California Civil Code', 65 *Calif. L. Rev.* 4 (1977).

[78] See *Muskopf* v. *Corning Hospital Dist.*, *supra*, at 216–17.

[79] See *Bing* v. *Thunig*, 2 NY 2d 656, 143 NE 2d 3 (1957).

even-handedness, prior support, replicability, the protection of justifiable reliance, and the prevention of unfair surprise— would be no better served by preserving the doctrine than by overruling it.[80]

A subsidiary point often raised in the debate over the legitimacy of judicial lawmaking concerns the weight, if any, to be given to evidence of any prior legislative attitude on the question involved. At one extreme, is the old saw that the legislature's very failure to intervene spells tacit approval of the existing judicial rule; but this unrealistically casts legislators into a false role, that of assiduous court watchers. Nor, for that matter, can one fairly read much into the fact that some statutes happen to assume the existence of the common law rule under attack: for obvious reasons, legislators in such instances simply did not focus attention on the merits of the rule itself.[81] Somewhat more troublesome would be evidence of recent contemporaneous legislative preoccupation with the question of reforming the common law rule. This might argue for giving the legislature more time rather than anticipating it.[82] On the other hand, repeated failure to enact legislation could also be interpreted as evidence of legislative inability to come to grips with the problem.[83]

But if the court's motivation is to relieve the legislature of

---

[80] Eisenberg, M., *The Theory of Adjudication*, ch. 7 (1988). The author specifically illustrates his thesis by the overruling of charitable immunity. The classical example is Cardozo's opinion in *McPherson* v. *Buick Motor Co.*, 217 NY 382, 111 NE 1050 (1916), disposing of the maze of inconsistencies and accumulated exceptions to the tort immunity of negligent manufacturers (*infra*, at n. 108). See Levi, *Introduction to Legal Reasoning* 8–27 (1949).

[81] 'Nor are we faced with a comprehensive legislative enactment designed to cover a field. What is before us is a series of sporadic statutes, each operating on a separate area of governmental immunity where its evil was felt most.' *Muskopf* v. *Corning Hosp. Dist.*, *supra*, at 218, 359 P. 2d at 461.

[82] This was one of the reasons why the Wisconsin Court refused to adopt the 'pure' comparative fault version in place of the statutory 'less than the defendant's' formula in *Vincent* v. *Pabst Brewing Co.*, 47 Wis. 2d 120, 177 NW 2d 513 (1970).

[83] Thus the Illinois Court in *Alvis* v. *Ribar*, *supra*, at 22–3, 421 NE 2d at 895–6, was not nonplussed by the previous introduction of comparative negligence bills; their failure was not thought to reflect a desire to retain the status quo, but rather to indicate a legislative judgment to leave the issue to the courts!

the task by anticipating what it presumably might have done itself, should it not heed any signs of legislative intent? Take, for example, the choice between 'pure' comparative negligence, i.e. apportionment in accordance with the respective shares of fault, and more limited versions such as the requirement that the plaintiff's fault be no greater than the defendant's. Almost all American statutes prefer the latter qualification,[84] advocated by the defence lobby,[85] and some courts introducing comparative negligence have heeded this signal.[86] Others, however, have ignored not only the general trend but even statutes in their own state, like statutes dealing with work accidents that were so qualified.[87] Such an attitude is more redolent of 'knowing better than the legislature' than pretending to be its surrogate.

Finally, there is the administrative factor. A strong argument against judicial intervention exists where the reform cannot be accomplished in one stroke but must instead await resolution of numerous details and collateral issues in its wake. Legislation would in such a setting be capable of disposing of most or all of these at once, issuing a complete comprehensive set of new regulations like Athena emerging full-grown from the forehead of Zeus. By this token, for example, governmental immunity seems to be less well qualified for judicial lawmaking compared with, say, the recognition of claims for wrongful birth[88] or the extension of consortium claims to both spouses.[89] Not that reform statutes invariably exploit this advantage—the New York statute abolishing sovereign immunity[90] is a notorious example of legislative irresponsibility, leaving all the details, including some important policy questions (e.g. discretionary

[84] See the list in *Alvin* v. *Ribar, supra*.

[85] *Responsible Reform: An Update* 15 (1972) recommending the even more prejudicial requirement, adopted by some states, that the plaintiff's fault be 'not as great as' the defendant's. Cf. *supra*, n. 82.

[86] *Langley* v. *Boyter, supra*.

[87] e.g. *Li* v. *Yellow Cab, supra*, ignoring Labor Code § 2801.

[88] Prosser and Keeton, *supra*, at 370–3; Harper, James, and Gray, *supra*, § 18.3–4.

[89] Prosser and Keeton, *supra*, at 931; Harper, James, and Gray, *supra*, § 8.9.

[90] Court of Claims Act § 8.

activities), to be worked out by the courts through the protracted process of random litigation.[91]

To a majority of commentators, at the time, the introduction of comparative negligence seemed to fall halfway between these extremes.[92] True, it would raise a number of possible collateral changes, but these seemed neither numerous nor complex enough to affect the balance critically. Two other considerations further mitigated their impact. First, almost all comparative negligence statutes were also in the briefest conceivable form, leaving the very same ancillary questions likewise to the courts for future solution.[93] Secondly, courts had often anticipated at least some of these and thus disposed of the need for having them explored later at the cost of future litigants.[94]

Later experience, however, has given pause. In California, for example, a veritable log-jam developed in the court system for several years, awaiting resolution by the Supreme Court of a number of commonly occurring problems. These included the question whether the mandate that responsibility should henceforth be commensurate with fault required abolition of the 'joint and several liability' rule and its compatibility with the contribution statute. How the court disposed of this challenge, which brought it into direct conflict with another inconvenient statute, will be postponed for later discussion.[95] Suffice it to say, however, that later legislative efforts to deal with these same problems over a protracted period have ended in complete failure due to the prevailing deadlock between the two professional interest groups.

[91] e.g. *Weiss* v. *Fote*, 7 NY 2d 579, 167 NE 2d 63 (1960) (4-second interval for amber traffic lights, fixed by Board of Safety, not justiciable).

[92] In a symposium, 'Comments on *Maki* v. *Frelk*—Comparative v. Contributory Negligence: Should the Court or Legislature Decide?', 21 *Vand. L. Rev.* 889 (1968) of 6 commentators only 1 argued against the court. To my embarrassment, my own discussion of the issue in 'Foreword: Comparative Negligence at Last: By Judicial Choice', 64 *Calif. L. Rev.* 289 (1976) has been invoked by courts justifying their initiative: e.g. *Placek* v. *City of Sterling Heights*, 405 Mich. 638, 657, 275 NW 2d 511, 518; *Alvis* v. *Ribar supra*, at 19-20, 421 NE 2d at 895-6.

[93] See Schwartz, *supra*, ch. 1.

[94] For example, the *Li* court expressed itself clearly against the survival of 'last clear chance' and less clearly against 'voluntary assumption of risk'.

[95] *Infra*, ch. 3, at n. 69.

3.3   Road Traffic Accidents

One of the more intriguing puzzles is that American courts have not tried to emulate the widespread Continental option of strict liability for motoring accidents. While this trend was mostly legislative elsewhere, starting with the German act of 1909, in France the reform was accomplished by the courts in a famous demonstration of judicial creativity without parallel in European annals.[96] In contrast, the abstention by American courts is the more remarkable in the light of their proven record of activism and their general support of the trend towards strict or stricter liability. One explanation is that this ideological shift came too late for a reform of automobile liability. At an early stage of motor traffic, arguments were advanced in the United States as they were in England to treat the 'devil wagon' as an ultra-hazardous object,[97] subsumable under the rule of *Rylands* v. *Fletcher* or, as it is now labelled by the *Restatement of Torts, Second*,[98] strict liability for 'abnormally dangerous activities'. These arguments were the more plausible where the motor car was defective, but they were voiced at a time of judicial stagnation, and the opportunity for reform was missed.

When the argument was revived, reinforced by the embarrassing analogy of strict products liability that had just been pushed through by the courts, it seems to have come too late. The California reaction is the more revealing for coming from a court in the forefront of judicial activism. In *Maloney* v. *Rath*,[99] the appellant urged that the defendant's violation of a safety provision in the Vehicle Code, requiring effective brakes, made the violator strictly liable when her brakes

[96] *Affaire Jand'heur*, Cass. chambres réunis, 13 Feb. 1930; Dalloz 1930.1.57. See my comment, 'Operating Defective Automobiles: The French Code and its North American Cousins', 23 *Am. J. Comp. L.* 513 (1975); Zweigert and Kötz, *An Introduction to Comparative Law* ii 322–9 (1977). A recent statutory reform has extended benefits to the culpable driver as well (infra ch. 5, note 90).

[97] See Ehrenzweig, *Negligence Without Fault* 22 (1951), citing *Lewis* v. *Amorous*, 3 Ga. App. 50, 55, 59 SE 338, 340 (1907), analogizing the automobile to a wild beast.

[98] §519. The first Restatement used the term 'ultrahazardous'.

[99] 69 Cal. 2d 442, 445 P. 2d 513 (1968).

unexpectedly failed, hitting the back of a car waiting at a stop light. The court, however, reaffirmed an earlier decision in which it had qualified the conventional negligence *per se* rule for statutory violation by excusing a violator who 'did what might reasonably be expected of a person of ordinary prudence, acting under similar circumstances, *who desired to comply with the law*'.[100] Its reason was plainly to avoid statutory negligence from becoming transformed from a test of negligence to a stratagem for no-fault liability. This would excuse unwitting violators who could not 'help it', as when a rear light suddenly goes out, brakes fail, or a tyre bursts and the driver loses control. Evidently, not even the circumstance that a strict application of negligence *per se* would here have promoted no-fault liability specifically for traffic accidents proved persuasive. Indeed, just the contrary. Chief Justice Traynor, rather more cautious than his radical posthumous admirers would have us believe, explained:

To invoke a rule of strict liability to users of the streets and highways, however, without also establishing in substantial detail how the new rule should operate would only contribute confusion to the automobile accident problem. Settlement and claims adjustment procedures would become chaotic until the new rules were worked out on a case-by-case basis, and the hardships of delayed compensation would be seriously intensified. Only the Legislature, if it deems it wise to do so, can avoid such difficulties by enacting a comprehensive plan for the compensation of automobile victims in place of or in addition to the law of negligence.[101]

Surprisingly, though, the appellant did succeed on another argument. For the court ruled that the statutory duty to drive with effective brakes was non-delegable and that the driver-owner was therefore liable for the negligence of the service station in overhauling her brake system. This form of vicarious liability, it said, was justified because driving was an 'activity which threatens a grave risk of serious bodily harm or death'[102]—an admission that might plausibly have qualified it independently for strict liability for 'abnormally dangerous

100 *Alarid* v. *Vanier*, 50 Cal. 2d 617, 622; 327 P. 2d 897, 899 (1958). Since given statutory authority by Evidence Code § 669.

101 *Maloney, supra*, n. 99, at 446, 445 P. 2d at 515.

102 *Restatement, Torts, Second* § 423.

activities'.[103] Yet this apparent incongruity did not overcome the court's disinclination to extend strict liability to all road traffic accidents.

Its avowed reason for this skittishness is not altogether convincing. Difficult adjustments during a protracted transitional period did not deter the introduction of comparative fault or strict products liability.[104] Perhaps closer to the truth is that no-fault compensation for motor traffic accidents had moved to the forefront of the legislative agenda and become a contentious public issue. While any additional cost of products liability would become a charge to industry and only eventually trickle down to consumers, any reform of motor vehicle liability would impinge directly on the public in the form of higher insurance premiums.

Indeed, the most contentious issue in the then current debate over automobile no-fault was precisely costs. It was the proposals to limit tort recovery (in order to achieve the necessary savings to offset the wider benefits) that aroused the vehement opposition of the trial bar, on which the scheme ultimately foundered in many states, including California.[105] All the more obviously, it would have been impossible for the courts to design a scheme containing similar compromises, if it had thought them desirable. Compromise is the life blood of democratic government but not of judicial adjudication in the common law.

3.4   Products liability

To many observers the most dramatic innovation of American tort law is the strict liability of manufacturers and distributors

[103] The court sought to explain the difference by saying that 'unlike strict liability, a nondelegable duty operates, not as a substitute for liability based on negligence, but ... to insure that there will be a financially responsible defendant available to compensate for the negligent harms caused by that defendant's activity ... [To that extent] it ameliorates the need for strict liability to secure compensation' (at 446, 445 P. 2d at 515).

[104] Another attempt to explain the difference in treatment of products and automobile liability on the ground that a producing and marketing enterprise should bear the cost of injury from defective parts is even more question-begging. See *Hammontree* v. *Jenner* 20 Cal. App. 3d 528, 532, 97 Cal. Reptr. 739, 741-2 (1971).

[105] *Infra*, ch 5, at n. 71.

of products introduced in the 1960s. In the shift from negligence to strict liability it resembled the daring of the French courts who construed Article 1384 of the *Code Civil* as a mandate of strict liability for the *guardien des choses* in the last decade of the nineteenth century and eventually extended that principle to drivers of motor vehicles in 1929.[106] It resembled the French *tour de force* also in affecting a vital and pervasive social activity with considerable impact on the economy. But this sense of its significance is more a judgment based on hindsight than the impression at the time of crossing the threshold. Moreover, while the courts and judges who played a key role in this development have been admired for their creativity, neither then nor since was the legitimacy, as distinct from the wisdom, of this judicial reform called into question or felt to be in need of justification.

The clue to this puzzling (non-)reaction is that the transformation of products liability occurred by stealth, over a long process of gradual evolution, so that its final consummation was seen as a logical progression rather than as an abrupt change of course. That its full implications for industry were not understood at the time is revealed by the virtual absence of any vehement protest. Indeed it was only the gradual unfolding of its evermore far-reaching implications over the following decade that moved products liability into the arena of contentiousness and stirred industry into active opposition.

A brief outline of this development may be helpful. The English decision of *Winterbottom* v. *Wright*[107] in 1842 first set the course of Anglo-American law, being understood as generally denying liability for a negligently defective product to persons other than the immediate purchaser. That this categorical insistence on 'privity' in delimiting the range of responsibility by negligent manufacturers and distributors did not go unquestioned in the course of time was demonstrated by the admission of an exception for inherently dangerous products—a category of ambiguous reference which gradually brought the primary rule into disrepute, as arbitrary dis-

---

[106] *Supra*, n. 96.

[107] 10 M. & W. 109 (1842). One recalls Bramwell, B's apocalyptic prediction 'of the most absurd and outrageous consequences to which I see no limits' if the rule were otherwise.

tinctions multiplied to the point where its eventual abrogation seemed overdue as an act of restoring rationality. This threshold is identified in American law with Judge Cardozo's decision in 1915 in *MacPherson* v. *Buick Motor Co.*,[108] which had its British counterpart in the 1932 House of Lords decision of *Donoghue* v. *Stevenson*.[109] It may be a sign of how long overdue was this belated recognition of negligence liability for defective products, no less than of the speed of social change since, that there was so little delay before its basis was transformed from negligence to strict liability in our own time.

This apparently volatile development was however aided by important doctrinal supports. One was the development of implied warranties of quality incident to the sale of goods. While in origin sounding in tort, warranties had by the end of the eighteenth century become subsumed under the law of contract; liability was strict but subject to the strait-jacket of privity.[110] As its name implied, the warranty of merchantability conveyed the original notion that the goods as delivered had to be merchantable, i.e. saleable, and it was confined to sales by description by merchants. In the course of time, however, the remedy became transformed by extension from commerce to consumer protection, as the result of providing a remedy not only for commercial loss but also for personal injury.[111] Moreover, sales by description were no longer confined to sales by correspondence but came to include face-to-face transactions.[112] This development had two effects. First, it familiarized the market with strict liability for defective products—incongruously, long before the law of torts required manufacturers to answer for their negligence. But secondly,

[108] 217 NY 382, 111 NE 1050 (1916). Cf. *supra*, at n. 80.

[109] [1932] AC 562.

[110] *Stuart* v. *Wilkins* (1778) 1 Doug. 18. See *Williston on Sales*, ch. 8 (rev. edn. 1948).

[111] See *Williston on Sales* (*supra*) § 614 consequential damages included in 'loss directly and naturally resulting, in the ordinary course of events, from the breach of warranty' (Uniform Sales Act § 69(6)). e.g. *Wren* v. *Holt*, [1903] 1 KB 610 (arsenic in beer).

[112] e.g. *Ryan* v. *Progressive Grocery Stores*, 255 NY 388, 175 NE 105 (1931); Prosser, 'The Implied Warranty of Merchantable Quality', 27 *Minn. L. Rev.* 117, 139–45 (1943). The qualification has been omitted from the Uniform Commercial Code § 2-314.

the benefit of this strict liability was confined to the immediate buyer of the goods. Because of the more complex nature of the distribution process, the intervention of middlemen typically came to preclude a direct contractual relationship between manufacturer and the ultimate consumer. This entailed the paradox that the manufacturer was not liable to him for breach of warranty nor even for negligence until much later, but that the retailer was. The party responsible for the defect thus escaped, while the innocent incurred a stringent liability which he might conceivably pass up the vertical line of distribution by invoking the benefit of his own implied warranty.

Though inelegant and not without parallel in other legal systems, this arrangement might have provided adequate consumer protection but for several vital gaps.[113] One was the insistence on privity between the victim and the defendant. This excluded all those outside the chain of commercial distribution, such as the ultimate buyer's family, friends, and bystanders. More important yet was the freedom to exclude warranties and the courts' refusal to protect buyers from improvidently accepting contracts of adhesion containing unconscionable disclaimers.

In recognition of the need for greater consumer protection, American courts first began to display 'considerable ingenuity' in evolving theories to avoid the objection of lack of privity and extend warranties to the consumer.[114] Mostly, this development was associated with sales of food and drink. It was aided in some cases by invoking pure food legislation (statutory violation), in others by resuscitating the memory of medieval guild regulations in support of a lasting historical commitment to public protection against adulterated food.[115] But as Prosser predicted as early as 1941 in the first edition of his famous text, this tendency

will be the law of the future, and . . . the end of the next quarter of a century will find the principle generally accepted. No reason is

[113] There were others besides those cited in this paragraph, e.g. under American sales law, the requirement of prompt notice to the seller.

[114] Prosser, *The Law of Torts* 691 (1941).

[115] Prosser, 'The Assault Upon the Citadel (Strict Liability to the Consumer)', 69 *Yale L. J.* 1099, 1103–10 (1960).

apparent for limiting it to food cases, and it may be anticipated that it will extend, first to other products involving a high degree of risk, and perhaps eventually to anything that may be expected to do harm if it is defective. If that is to occur, it seems far better to discard the troublesome sales doctrine of 'warranty,' and impose strict liability outright in tort, as a pure matter of social policy.[116]

True to this prediction, the category of food and drink was first expanded to include other ingestibles like vaccines[117] and cigarettes.[118] From 1960 the quickening development can be traced in successive drafts of *Restatement of Torts, Second* which was then being prepared by the American Law Institute. Section 402A, as originally submitted in 1961 was still limited to food for human consumption.[119] The following year the Reporter, Dean William Prosser, had expanded it to include 'products for intimate bodily use'[120] in the light of decisions implying warranties *sans* privity from cosmetics to hula skirts. Finally, in 1964 he proposed the current version of 'any product in a defective condition unreasonably dangerous to the user or consumer or his property'.[121] He noted 'so many decisions extending the strict liability beyond products "for intimate personal use" that it has become quite evident that this is the law of the immediate future'.[122] This prognosis became a self-fulfilling prophecy as the authority of the Institute became itself a potent argument for adopting its black-letter.

In 1966 Prosser celebrated the 'Fall of the Citadel' of privity, noting that it represented 'the most rapid and altogether spectacular overturn of an established rule in the entire history of the law of torts'.[123] For him, the New Jersey decision in

[116] Prosser, *supra*, n. 114, at 692.

[117] *Gottsdanker* v. *Cutter Laboratories*, 182 Cal. App. 2d 602, 6 Cal. Rptr. 320 (1960).

[118] *Green* v. *American Tobacco Co.*, 154 So. 2d 169 (Fla. 1963).

[119] Tent. Draft No. 6.

[120] Tent. Draft No. 7.

[121] Tent. Draft No. 10.

[122] Ibid.

[123] 'The Fall of the Citadel, (Strict Liability to the Consumer)', 50 *Minn. L. Rev.* 791, 793-4 (1966). Note Prosser's earlier article, *supra*, n. 115. The reference to the 'citadel' harks back to Judge Cardozo's famous dictum in *Ultramares Corp.* v. *Touche*, 255 NY 170, 180; 174 NE 441, 445: 'The assault upon the citadel of privity is proceeding in these days apace.'

Henningsen v. *Bloomfield Motors*[124] had effected the breach, since widened by seventeen other jurisdictions. For Californians, it is linked to a decision fifteen years earlier when Justice Traynor[125] had lent his rising authority to the view that, instead of resorting to the stratagem of *res ipsa loquitur*, it was time to declare outright that public policy demanded that the manufacturer stand behind his product and guarantee its safety in the interest of accident prevention and consumer protection.[126] It was a personal triumph for Traynor to enunciate this doctrine on behalf of a unanimous court in *Greenman* v. *Yuba Power Products* in 1963.[127]

Thus far the innovation was seen more as a cleaning-up operation, freeing manufacturers' liability merely from the 'intricacies of the law of sales'[128] and dispensing with the need to stretch *res ipsa loquitur*, rather than as opening up a novel and potentially explosive source of liability. Indeed, one of its mooted attractions was that it would reduce litigation by stopping up undeserved loopholes through which a guilty defendant might hope to escape. Unhappily, these sanguine expectations were soon turned to ashes as it became clearer that the courts were opening a Pandora's box. Products liability was to transform the whole agenda of tort litigation: inaugurating a new era of mass actions, mass costs, and mass awards.[129]

[124] 32 NJ 358, 161 A. 2d 69 (1960).

[125] On Traynor's contribution to torts see Malone, 'Contrasting Images of Torts: The Judicial Personality of Justice Traynor', 13 *Stan. L. Rev.* 779 (1961); White, *Tort Law in America*, ch. 13 (1980).

[126] *Escola* v. *Coca Cola Bottling Co.*, 24 Cal. 2d 453, 150 P. 2d 436 (1944).

[127] 59 Cal. 2d 57, 377 P. 2d 897 (1963). Ever since, the California Court has gloried in this feat: e.g. 'Our pioneering efforts in this field [products liability]' (*Cronin* v. *J. B. E. Olson Corp.*, 8 Cal. 3d 121, 133, 501 P. 2d 1153 (1972)); '[t]his Court heroically took the lead': *Daly* v. *General Motors Corp.*, 20 Cal. 3d 725, 757, 575 P. 2d 1162 (1978). The implications of Traynor's reasoning on the general direction of tort law were considered *supra*, ch. 1, at n. 35.

[128] *Ketterer* v. *Armour & Co.*, 200 Fed. 322, 323 (SDNY 1912).

[129] *Supra*, ch. 1, at n. 44.

3.5 Design defects

The trouble arose over an issue seemingly self-explanatory: when is a product defective? To placate conservative critics the point has been frequently made that the new liability was not absolute like insurance: the manufacturer was not liable for every accident resulting from use of his products, only for accidents caused by a defect. *Restatement of Torts, Second* itself bears testimony that the stock situation contemplated by the Rule was a fault in the production process, where only the plaintiff's lack of access to the evidence prevented him from identifying exactly what had gone wrong. Nobody would dispute that new cars should not be equipped with faulty brakes or steering, that Coca Cola bottles should not contain wayward mice or explode in your face, or that lawn-mowers should not fling their blades into the user's legs. All these are what are commonly called manufacturing defects. Their characteristic is that the defective product deviates from the manufacturer's own specifications, which themselves provide the appropriate standard for judging the quality of the product.

The situation is very different in the case of so-called design defects. Here the product conforms to the manufacturer's design, but the design itself is attacked as substandard. The *Restatement* failed to draw this distinction, which was still awaiting exfoliation, but recognized the collateral problem of 'unavoidably unsafe products', especially common in the field of drugs. In order to exclude these from strict liability because they were useful and desirable, the *Restatement* insisted that the Rule applied only to 'unreasonably dangerous products', not to those containing a known but apparently reasonable risk.[130] But how are we to determine when a product that theoretically could be made safer is not safe enough and therefore arguably unreasonably dangerous? Design defect cases had occasionally been litigated earlier but without realizing the distinction. Easiest was the case of the seller's

---

[130] § 402A, Comment *k*. See also Comment *i*. But this would not necessarily exclude a duty to warn.

assurance that the car's windscreen was 'shatter-proof'.[131] He was held liable to the passenger for breach of express warranty, but the standard was set by the promise. In its absence, however, how would one have decided in 1932 whether the windscreen should have been shatter-proof? Another type of case also self-answers this question, namely that where the alleged defect defeats the very purpose which the product is designed to serve and thereby causes the accident. Examples are a carburettor which, as designed, tends to stick, a jack that collapses under the weight of its load, or a lathe that does not securely grip a piece of wood. In these examples the design flaw was not intended but due to inadvertence.

The real difficulty concerns design features resulting from a calculated choice by the designer. That choice is usually dictated by a combination of various factors in which safety and cost tend to be prominent but not exclusive. The difficulty is aggravated where the particular feature did not cause the accident but only aggravated its consequences, as when a motor car should arguably have been 'crash-proof'. A heavier steel construction would, no doubt, give better protection than a light skin but at a cost of reduced speed, greater fuel consumption, and higher price. Thus safety is here opposed by consumer preference and national energy policy. Such decisions are traditionally made by design engineers (and their boardroom superiors); alternatively they could be made by independent bodies, such as trade organizations or public agencies. An example of the latter is the National Highway Traffic Safety Administration in the United States, which has been responsible for a number of mandatory safety features, including safety harnesses, head rests, tempered safety glass, and others. The drawn-out regulatory controversy over airbags illustrates the sensitivity of some of these issues. The critical question for tort liability is whether such decisions should be entrusted to judge and jury.[132]

[131] *Baxter* v. *Ford Motor Co.*, 168 Wash. 456, 12 P. 2d 409 (1932). The rule is embodied in *Restatement, Torts, Second* § 402B and in Uniform Products Liability Act § 104.

[132] Where regulatory supervision is comprehensive and trustworthy, as in the case of drugs, there is a strong case for treating FDA approval as precluding a design defect claim: *Collins* v. *Karoll*, 186 Cal. App. 3d 1194, 231 Cal. Rptr. 396 (1986).

Advocates for the affirmative case argue that calculated design choices are the result of deliberate business decisions linking safety investment to profit margins and as such peculiarly suitable for control through potential tort liability. Moreover, such decisions are qualitatively no different from the usual issues in negligence litigation, involving what for short may be called a cost–benefit analysis. Thus it is easy enough to brand as indefensible a managerial decision to dispense with a safety device for the sake of saving a negligible sum of money, like the extra $2 for a mixing valve that would have saved a child from being scalded by boiling water from a shower rose.[133] But few managerial choices are so obviously flawed. More usually there is no agreed criterion for resolving what Professor Fuller has called a 'polycentric' problem, i.e. a many-centred problem in which each point for decision is related to all the others as are the strands of a spider's web.[134] 'Intelligent answers to the question of "how much product safety is enough?" . . . can only be provided by a process that considers such factors as market price, functional utility, and esthetics, as well as safety, and achieves the proper balance among them. Ultimately the question reduces to "What portion of society's limited resources are to be allocated to safety, thereby leaving less to be devoted to other social objectives?" '[135] The persisting struggle by American courts to devise a suitable definition of 'defectiveness' is symptomatic of this problem.[136] Justice Traynor himself came to acknowledge that none have proved adequate to define the scope of the manufacturer's strict liability.[137]

[133] *Schipper* v. *Levitt & Sons, Inc.*, 44 NJ 70, 207 A. 2d 314 (1965).

[134] See especially Fuller, 'Forms and Limits of Adjudication', 92 *Harv. L. Rev.* 353 (1978). On Fuller's theory see Summers, *Lon L. Fuller* 98–100 (1984).

[135] Henderson, 'Judicial Review of Manufacturers' Conscious Design Choices: The Limits of Adjudication', 73 *Colum. L. Rev.* 1531, 1540 (1973).

[136] The California court, for example, abandoned the *Restatement* formula of 'unreasonably dangerous' for 'burden[ing] the plaintiff with proof of an element which rings of negligence', only to revert later to the familiar balance calculus of the negligence test, but with a reversal of the onus of proof as the sole token of strict[er] liability: *Barker* v. *Lull Engineering Co.*, 20 Cal. 3d 413, 573 P. 2d 443 (1978). See Prosser and Keeton, *supra*, 698–702.

[137] 'The Ways and Meanings of Defective Products and Strict Liability', 32 *Tenn. L. Rev* 363, 373 (1965).

Professor James Henderson has accordingly argued that the process of adjudication administered by the courts is not institutionally suited to establish product safety standards.[138] Every alternative advanced by either party would have to be evaluated in the light of many factors, of which safety is but one. The judicial process is accustomed to apply specific standards established *aliunde* by decisional processes more suited to the task, but it is unsuited to establish such standards itself. It would follow, he submits, that

a broad-scale judicial commitment to the independent review of conscious design choices would bring with it a very real threat to the integrity of the judicial process. Confronted with the hopeless difficulties of trying to redesign products via adjudication, and presumably unable to resist the social pressures generally favoring injured plaintiffs, courts would inevitably resort to some form of judicial coin-flipping, i.e. they would begin to determine defendants' liability on some arbitrary basis . . . In effect, the adjudicative process would largely become a sham. Although such tactics might render these cases manageable in the short run, they would do so at the cost of a serious erosion of confidence in the courts by those litigants who would correctly come to realize that they have been denied effective access to the adjudicative process by such subterfuge.[139]

This criticism suggests that, in extending the law of products liability to conscious design choices, the courts have been lured into assuming a role for which they lack both competence and legitimacy, alongside such other questionable ventures as the judicial management of schools and prisons.[140] Indeed, far from strict liability reducing litigation, the volume of claims has exploded to a point where it is widely blamed for the increasing log-jam in the courts. 'Old-fashioned' claims for manufacturing defects are disposed of by settlement, unless causation or damages are in dispute. Litigation is mostly over design defects. Such claims are bitterly fought by defendants

[138] *Supra*, n. 135.
[139] Ibid., at 558.
[140] Sanctioned by the US Supreme Court in *Swann* v. *Charlotte–Mecklenburg Board of Education*, 402 US 1 (1971) (judicially constructed affirmative desegregation plan). See Chayes, 'Foreword: Public Law Litigation and the Burger Court', 96 *Harv. L. Rev.* 4 (1982); D. Horowitz, *The Courts and Social Policy* (1977).

because of the peculiarly prejudicial effects of adverse judgments. An adverse judgment does not condemn merely one particular item that has strayed from its norm, as in the case of a manufacturing defect; it condemns a whole production line. Even if not necessarily binding on other claims,[141] it sets the tone and often receives wide publicity which will be reflected in loss of sales, as well as in a tendency to influence settlements and juries in following cases.

'Failure to warn' has become, if anything, an even more volatile source of liability because it does not necessitate an attack on the quality of the product itself but merely on the manner of its distribution. Moreover, the burden of preventive measures is relatively slight and therefore favours liability for failure to use them. Both types of case, however, raise troubling questions as to whether foresight or hindsight should control the question whether the defendant had or should have had knowledge of the risk involved. The New Jersey Court once adopted the extreme (and intellectually provocative) view that a manufacturer is liable for failure to warn of risks although unknown and unknowable at the relevant time.[142] Other courts have, however, shrunk from so far divorcing products liability from prevailing notions of responsibility.[143]

Other significant departures from the traditional boundaries of tort liability are involved in procedural adjustments for mass litigation. One of these is the promotion of class actions;[144] another is the flirtation with 'industry-wide' liability in order to assist plaintiffs who cannot identify particular culprits from among a group.[145] Both developments are markers on the road from individual responsibility to collectivism, moving tort law from the familiar function of 'commutative' to that of 'distributive' justice. For courts this direction is radical in two respects: first, in deciding upon this

[141] *Res judicata* and collateral estoppel are discussed *infra*, ch. 7, at n. 40.

[142] *Cepeda* v. *Cumberland Engineering Co.*, 76 NJ 152, 386 A. 2d 816 (1978); held inapplicable to drugs in *Feldman* v. *Lederle Laboratories*, 97 NJ 429, 479 A 2d 374 (1984).

[143] 'This is a minority view and can reasonably be regarded as an extreme position': Prosser and Keeton, *supra*, 701; Henderson, 'Coping With the Time Dimension in Products Liability', 69 *Calif. L. Rev.* 919 (1981).

[144] Infra, ch. 7.

[145] Ibid.

redirection which is more germane to the legislative than the judicial function; secondly, in taking the law of torts off its hinges and transforming it into an instrument for wealth redistribution.

3

CONSTITUTIONALITY

The relationship between courts and legislature has obvious constitutional implications. Being conditioned by current political perceptions, it is apt to vary with time and place. In most countries, even those priding themselves on an in-dependent judiciary as a badge of legality, governments (executive and legislature) are jealous of pretensions by the judiciary to make rather than to merely administer law. In Western democracies, the subordination of courts to the legislature has a venerable political tradition, reinforced in most of them by the absence of any entrenched constitution which would make the courts, or at least a constitutional court, the final arbiter of powers exercised by the other branches of government.

American perceptions about that relationship are very different. Under the federal and state constitutions, the judiciary is a co-ordinate branch of government, of equal and independent status to that of the executive and legislature. While the exercise of that power is most visible on constitutional issues, it is pervasive throughout the courts' whole jurisdiction. That constitutional issues are litigated at all levels of the judicial system, in contrast to the modern European pattern of a single Constitutional Court, contributes to the perception that the courts have the last word on the exercise of all legislative and executive power. Since the Second World War the focus of federal constitutional law in the United States has noticeably shifted from state–federal relations to the Bill of Rights, resulting in a wholesale extension of civil rights *vis-à-vis* state as well as federal government and in an explosive enlargement of those rights, especially those of minorities. Courts have consequentially come to see themselves more and more as protectors of the individual against abusive conduct by government and other powerful institutions. In the torts field this has reinforced other tendencies favouring plaintiffs against corporate and institutional defendants.

The fall-out of this new orientation of judicial attitudes to legislation has several facets. In the first place, it has contributed to the notorious 'judicial activism', discussed in the previous chapter, which rests on the view that legislation has no monopoly on bringing about legal change. The generally timid legislative response has tended only to encourage judicial confidence in the judges' new found role as law reformers. Secondly, conservative legislation, especially that emanating from defence interests, has become an inviting target of judicial scrutiny. The assault has proceeded both on constitutional grounds and by sleight of hand. Both techniques have yielded fascinating vignettes of judicial sophistry.

1. CONSTITUTIONAL ADJUDICATION

Tort statutes may be vulnerable to constitutional attack on several grounds: under the federal Constitution for denial of equal protection or due process; for their counterparts under state constitutions or for other constitutional infirmities, special to the state, such as guarantees of separation of powers or access to courts. Contrary to what one might perhaps have expected, state courts have on the whole evinced less compunction about striking down their own statutes than have federal courts in ruling on the constitutionality of state statutes.[1] Moreover, it has become fashionable for reform-minded courts to rely on their own state constitution rather than on the federal Constitution to give a more liberal construction than the US Supreme Court's to such open ended concepts as equal protection and due process and thereby render their ruling proof against reversal on appeal.[2] This

---

[1] In applying both federal and state constitutions. An important difference is that federal courts pay some deference to state laws out of concerns for federalism. Based on a decade's experience, ATLA recently identified state constitutions rather than the federal Constitution as the best way to challenge tort legislation: *ATLA Advocate*, vol. xii, no. 7, p. 1.

[2] The revival of interest in state constitutions is a phenomenon barely a decade old. Inconsistent interpretations of identically worded constitutional

explains to some extent the lack of unanimity concerning the relevant constitutional tests, which often mirror, in the present context, a difference of ideology concerning the proper role of tort law.

1.1 Equal protection

The Fourteenth Amendment prohibits a state from denying 'to any person within its jurisdiction the equal protection of the laws'. Tort reform statutes in the American tradition tend to deal with specific trouble-spots rather than with broad categories. By engrafting exceptions to a general common law principle, such as liability for negligence, this narrow focus creates distinct classifications with resulting unequal treatment. Not that the Constitution forbids all discrimination. '[It] does not require that things different in fact be treated in law as though they were the same. But it does require, in its concern for equality, that those who are similarly situated be similarly treated. The measure of the reasonableness of a classification is the degree of its success in treating similarly those similarly situated.'[3]

The US Supreme Court has laid down two principal standards for testing conformity to this constitutional requirement: strict scrutiny and minimal rationality. Strict scrutiny is appropriate for so-called 'suspect' classifications, like race, religion, nationality, alienage, and for classifications involving 'fundamental rights' explicitly or implicitly guaranteed by the Constitution, like marriage and voting.[4] These

guarantees, based not on demonstrable variant local factors but motivated solely to give wider scope to civil rights, are the most controversial aspect of that 'new federalism'. See generally Collins, 'Foreword: Reliance on State Constitutions: Beyond the "New Federalism" ', 8 *U. Puget Sound L. Rev.*, no. 2, p. vi (1985); Williams, 'In the Supreme Court's Shadow: Legitimacy of State Rejection of Supreme Court Reasoning and Result', 35 *So. Ca. L. Rev.* 353 (1985).

[3] Tusman and tenBroek, 'The Equal Protection of the Laws', 37 *Calif. L. Rev.* 341 (1949).

[4] Most of these have been spelled out only in recent years, e.g. privacy (*Roe* v. *Wade*, 410 US 113 (1973) (abortion), right to travel (*Attorney General of New York* v. *Soto Lopez*, 106 S. Ct. 2317 (1986)), right of association (*NAACP* v. *Alabama*, 357 US 449 (1958)).

can be justified only by a 'compelling' state interest. Tort rules are unlikely to fall into this category.

Traditionally, other classifications have been relegated to the minimalist requirement that the challenged measure rationally promote legitimate governmental objectives.[5] 'Legitimate' means only that the purpose must not violate a specific constitutional prohibition. The test disclaims all concern with the wisdom of the measure and is so deferential that, if taken literally, it would rarely, if ever, support a successful challenge, since it is difficult to imagine legislation so irrational that it fails altogether to promote a given purpose.[6] However, in order to preserve more than a semblance of oversight against arbitrary and capricious classifications, modifications of the test have been suggested to make it conformable, or more conformable, with the course of some more intrusive state court decisions. One is the so-called 'means-focused' test which measures the rationality of the measure by the degree of success in furthering its objective.[7] It would be concerned with 'overbreadth' and 'under-inclusiveness', criteria which play a traditional role in strict scrutiny analysis. Another advocates a 'sliding scale' approach under which the degree of scrutiny varies with the 'constitutional and societal importance of the interest adversely affected and the recognized invidiousness of the basis on which the particular classification is drawn'.[8]

Both have links to an intermediate standard, so far invoked by the US Supreme Court only for sex and legitimacy-based classifications, which must 'substantially further' a purported

[5] A late example is *Lyng* v. *Castillo*, 106 S. Ct. 2727 (1986) (upholding a definition of 'household' under the federal food stamp programme).

[6] The rationality test can, and has been, ridiculed for making of our courts 'lunacy commissions sitting in judgment upon the mental capacity of legislators and, occasionally, of judicial brethren' (Cohen, 'Transcendental Nonsense and the Functional Approach', 35 *Colum. L. Rev.* 809, 819 (1935)). But why should we strain to find a role for judicial review of non-suspect legislation?

[7] Gunther, 'Foreword: In Search of Evolving Doctrine on a Changing Court: A Model for a Newer Equal Protection', 86 *Harv. L. Rev.* 1 (1972).

[8] *San Antonio Independent School District* v. *Rodriguez*, 411 US 1, 98–9 (1973); *Lyng* v. *Castillo, supra*, at 2732–4 (continuing dissents by Marshall, J., its chief protagonist).

legislative purpose. Although the reviewing court pretends not to question the wisdom of the legislative objective, the state must give greater justification for the means adopted than is required for rational basis analysis.[9] An example in the tort context is the common law action for loss of consortium, which traditionally was available to the husband but not to the wife. This instance of invidious sex discrimination had to be rectified by either abolishing the husband's action or extending it also to the wife. Challenged both for its archaism and for constitutional violation, courts have resolved the problem either way.[10] Social reformers, recognizing the potential of this wedge, claim it also in aid of 'basic fairness to a disadvantaged group'. Sporadic judicial support has been given to including even tort plaintiffs, somewhat disingenuously ignoring their powerful representation by the organized trial bar.[11]

1.2   Guest statutes

The first major testing ground for constitutional attack on tort statutes turned out to be the widely contemned guest statutes. Enacted by a majority of states in the decade following 1927 at the behest of the insurance lobby, their purport was to reduce substantially the liability of motor-car drivers to guest passengers.[12] The California statute, typical of the genre, provided that

No person who as a guest accepts a ride in any vehicle upon a highway without giving compensation for such ride . . . has any right

[9] Extended to education in *Serrano* v. *Priest*, 18 Cal. 3d 728, 557 P. 2d 929 (1976). This decision, by barring discrimination between school districts by wealth, revolutionized California's system of school financing.

[10] On constitutional grounds: *Karczewski* v. *Baltimore & Ohio R. R. Co.*, 274 F. Supp. 169 (Ill. 1967); *Hastings* v. *James River Aerie N. 2337*, 246 NW 2d 747 (ND 1976). Generally Prosser and Keeton, *Law of Torts* 931-2 (5th edn. 1984); Harper, James, and Gray, *Law of Torts* § 8.9 (2nd edn. 1986). The California court tried both solutions: *Rodriguez* v. *Bethlehem Steel*, 12 Cal. 3d 382, 525 P. 2d 669 (1974).

[11] See *Brown* v. *Merlo*, *infra*, n. 15.

[12] See Comment, 'The Constitutionality of Automobile Guest Statutes: A Roadmap of the Recent Equal Protection Challenges', [1975] *BYU L. Rev.* 99. For Canadian counterparts see Linden, *Canadian Tort Law* (3rd edn. 1982) 613.

of action for civil damages against the driver on account of personal injury or death . . . [unless it] resulted from intoxication or willful misconduct of the driver.[13]

This legislation quickly forfeited judicial favour as courts became embroiled in fertile litigation over the meaning of 'accepting a ride without compensation' and 'willful misconduct' (or similar disqualifications such as recklessness, gross negligence, etc.). At one point, the California legislature intervened to resolve a much disputed question by adding at the beginning of the section, the words 'no owner or [person who . . .]', with the intent of treating car owners like all other passengers who paid no compensation to the driver. Judicial displeasure was widely expressed not only in artificially narrow constructions of the statute but in open condemnations of it as unfair and, eventually, obsolete. In the absence of any legislative response to these protests but emboldened by the judicial enthusiasm for civil rights in the Warren era,[14] a challenge to the California guest statute for violation of the Equal Protection clause of the Fourteenth Amendment was ultimately launched, culminating in its invalidation by the Supreme Court of California in 1973 in *Brown* v. *Merlo*.[15]

The challengers contended that the statute unjustifiably discriminated between automobile guests and other recipients of hospitality, between automobile guests and paying passengers, and between accidents on the highway and elsewhere. Did these distinctions bear a rational relationship to any permissible purpose(s) the legislature was seeking to promote? The US Supreme Court, far back in 1929, had upheld the Connecticut statute, refusing to inquire into its wisdom and repudiating any 'constitutional requirement that a regulation, in other respects permissible, must reach every class to which it might be applied'.[16] Thus the fact that it did not purport to affect guests in other forms of transportation was held not to be prohibitive. The California Supreme Court curtly

13 Vehicle Code § 17158 (enacted in 1929).
14 Earl Warren was Chief Justice of the United States 1953-69. See generally Choper, 'On the Warren Court and Judicial Review', 17 *Cath. U. L. Rev.* 20 (1967).
15 8 Cal. 3d 855, 506 P. 2d 212 (1973).
16 *Silver* v. *Silver*, 280 US 117 (1927).

brushed aside this decision on the ground that it did not pass on the other discriminations mentioned. Besides, the intervening forty years had seen several changes in the legal and factual setting of the guest statutes, such as compulsory liability insurance and the court's abolition of prejudicial treatment for 'licensees', permissive visitors such as social guests on the defendant's land.[17] It was not disputed that a measure can suffer constitutional obsolescence as a result of intervening changes, not only of constitutional doctrine but also of legal ambiance: rationality must be tested *nunc*, not *tunc*.

In the absence of a clear legislative expression of its objective, the California Court addressed two purposes 'traditionally advanced in both judicial precedents and academic commentaries': the protection of hosts from ungrateful guests, and the prevention of collusive law suits.

The hospitality rationale had been undermined, as already mentioned, by several changes of law: the abolition of charitable immunity and of the status distinction between the duty owed by occupiers to licensees and to other visitors. The rationale itself embodied two distinct strands of reasoning. First was the proposition that 'you get what you pay for'. Yet there was no general principle recognized by the common law or statute that one must pay a fee for being entitled to legal protection against negligent injury. A distinction between paying and non-paying passengers could rationally be drawn but not at the expense of withholding all protection against negligence from the latter group; for example, the common-law rule entitling passengers of commercial carriers to the 'utmost care' was quite defensible. The second strand was the 'Good Samaritan' argument that it was inexcusable ingratitude to sue your host. Liability insurance, however, had completely

---

[17] The California Court adopted the ordinary duty of care for all visitors in *Rowland* v. *Christian*, 69 Cal. 2d 108, 443 P. 2d 561 (1968). Note the bootstrapping: the court, by its own prior reform of the common law, was thus able to undermine the earlier statute! A similar stratagem was rejected in *Reda Pump Co.* v. *Fink*, 713 SW 2d 818, 821 (1986) when it was urged that a 1978 statute making contributory negligence a complete defence to products liability was invalidated by the judicial introduction of comparative negligence in 1974: 'To hold [otherwise] would constitute the ultimate arrogation of power unto this Court.'

nullified this contention since there is no ingratitude in suing your host's insurer. Finally, the distinction bore no relation to the realities of life since, the court postulated in the teeth of economic theory, reasonable people do not vary their conduct by such factors as the exchange of payment, nor does the reduced standard of liability encourage hospitality.

The second assumed purpose of the guest statute, fear of collusion between passenger and driver against the insurance company, was dismissed as 'overinclusive', i.e. the classification imposed a burden on a wider range of individuals included in the class of those tainted by the mischief at which the law aims. While the guest statute commonly evoked the image of the ingrate hitchhiker, in reality it was more likely to be that of family members.[18] These could not have sued the driver until the family immunity fell to the reforming axe of the court in the early 1960s;[19] since then, however, they had become the principal victims of the guest statute. But how could disqualification of all passengers be justified, considering that not all drivers would yield to the temptation of conspiring against the insurer, and in any event the risk of collusion was not confined to vehicular accidents?

Finally, limitations to accidents 'on the highway' or (under some guest statutes) 'during a ride' could not be related at all to either of the two assumed legislative purposes.

The foregoing analysis contained serious flaws which even the California Court subsequently admitted.[20] In its own words, it 'departed from the traditional equal protection standard' in several respects. First, by limiting itself to justifications of the statute traditionally advanced by judicial

[18] The Texas statute, upheld in *Tisko* v. *Harrison*, 500 SW 2d 565 (Tex. Ct. App. 1973) was expressly amended to include only relatives 'within the second degree of consanguinity or affinity'.

[19] *Emery* v. *Emery*, 45 Cal. 2d 421, 289 P. 2d 218 (1955). See *supra*, ch. 2, following n. 21. There too, collusion had been repudiated as sufficient support for retaining the family immunity, but that involved overruling a common law precedent, not invalidating a statute.

[20] *Schwalbe* v. *Jones*, 16 Cal. 3d 514, 518 n. 2 (1976), delicately hiding this confession of error in a footnote. It is also revealing that the *Merlo* court had cavalierly relegated to a footnote the US Supreme Court's decision in *Silver* v. *Silver*, *supra*, upholding the Conn. guest statute against a challenge of 'underinclusion'.

precedents and academic commentaries. Indeed, the principal purpose of the insurance industry in lobbying for guest statutes was, surely, to reduce premiums and thereby make insurance more affordable, obviously in itself a perfectly legitimate governmental purpose. Was it not permissible for the legislature to seek that objective by means of limiting the exposure of drivers' liability to passengers who make up about one-third of all injuries and are predominantly members of their family?[21] Arguably, this objective might also have been attained without changing the liability rules, for example by not requiring insurance cover for gratuitous passengers, but that would still have left drivers exposed to potential liability and to higher premiums for those opting for complete cover. More important still, the existence of alternatives (perhaps even of alternatives which arguably might have come closer to realizing the legislative objective) is not a recognized standard for constitutional review of minimal rationality. It would be questioning the wisdom of the legislation, which is universally disclaimed.

The court also regretted that 'the technique of expressing a suggested legislative purpose in terms of a "super class" (recipients of hospitality) and thereby proceeding to level the charge of underinclusion at the particular statute, might be interpreted to ignore the fundamental principle that the legislature, in dealing with what it deems to be an evil, is not held to a rigid choice of regulating all cases affected by that evil or none at all'.[22] This confession of error in effect invalidated most of the argument in *Merlo*, particularly with respect to collusion. Thus, although collusion is a risk in all claims between family members with a background of insurance, the problem may well have been thought to call for a special remedy in motoring cases where insurance is so widespread or even mandatory.

Lastly, the court expressed regret for having required that the statute bear a 'fair and substantial' relationship to an actual state purpose. While it had repudiated the contention

[21] In California at least 15% of all motor vehicles are not covered by third-party liability insurance, although it is mandatory. See *infra*, ch. 5, at n. 79.
[22] *Supra*, n. 20.

that the guest statute involved 'fundamental rights', let alone created 'suspect classifications' calling for strict scrutiny,[23] it posed in effect a stricter standard than the traditional minimalist 'rationality' test by inquiring into whether the statutory classifications were overbroad or underinclusive, whether the legislative classifications in fact furthered the assumed legislative purposes, and whether their relative importance justified the severity of the disabilities imposed on automobile guests.[24]

However, it was now too late to save the guest statute, for in the meantime the legislature had cleaned up the statute book by repealing all of the stricken section relating to guest passengers, but—perplexingly—retaining its application to automobile owners who permit themselves to be driven by others. The constitutionality of this rump came before the Court three times in the span of three years, first being declared invalid, then valid, and finally and definitively invalid (the vacillation being attributable primarily to shifting personnel, thereby illustrating another facet of the instability of American adjudication[25]).[26] Though the court had in the meantime intimated that the original decision in *Brown* v. *Merlo* had been based on a dilution of the traditional standard of constitutionality, which it now desired to disclaim, the corpse could not be revived after its *coup de grâce* by the legislature. There being no realistic independent ground for

---

[23] Ibid.

[24] Comment, *supra*, at 110.

[25] Another remarkable illustration is the well-known mental shock case of *Dillon* v. *Legg*, 68 Cal. 2d 728, 441 P. 2d 912 (1968), reversing *Amaya* v. *Home Ice, Fuel & Supply Co.*, 59 Cal. 2d 295, 379 P. 2d 513 (1963). The swing vote was by a judge, since promoted, who had been reversed in the first appeal. He also happened to have the last word in *Cooper* v. *Bray* (*infra*, n. 26). That judges on most courts of last resort feel at liberty to follow their own drummer rather than defer to collegiate solidarity is of course in striking contrast to British practice: see Paterson, 'Lord Reid's Unnoticed Legacy: A Jurisprudence of Overruling', 1 *Oxf. J. Leg. St.* 375 (1981).

[26] (1) *Schwalbe* v. *Jones*, 14 Cal. 3d 1, 534 P. 2d 73 (1975); (2) on rehearing, *Schwalbe* v. *Jones*, *supra*, n. 20; (3) *Cooper* v. *Bray*, 21 Cal. 3d 841, 582 P. 2d 604 (1978). The composition of the court had changed radically, converting the dissent by Tobriner, J., into the majority opinion delivered by him in *Cooper* v. *Bray*.

sustaining its vitality *vis-à-vis* vehicle owners,[27] the statute was declared void *in toto*.

Courts in a number of other jurisdictions were only too happy to follow the California precedent in order to rid themselves of the odious guest statute;[28] in a few others legislators belatedly repealed it.[29] This unhappy episode none the less taught a salutary lesson against irresponsible judicial adventurism. After freely reforming the common law to fit its own vision of contemporary values, the California and other like-minded courts lost their sense of perspective when trying to ride roughshod over statutes as they had done over venerable common-law precedents. Not only did this violate a deeply entrenched constitutional convention concerning the separation of powers; it also, by invoking the power of constitutional review, deprived the legislature of the opportunity to have the final word. When the court overrules its own precedents despite long standing, it sends a message to the legislature that the time for review is overdue, that it has set a new course which can, however, be reversed if not acceptable to the majoritarian branch of government.[30] The problem with obsolete statutes is less tractable.

Various devices are available for courts to nudge the legislature to take a second look at statutes that have become obsolete or suffer from constitutional infirmities. These range from the German constitutional device of setting a definite time limit for legislative amendment to bring a statute into constitutional conformity,[31] to such American doctrines as

---

[27] The legislative purpose must be real, not fictitious as adumbrated in the second appeal (*supra*, n. 26).

[28] See Comment, *supra*.

[29] Colo., Conn., Fla., Kan., Mont., Vt., Wash.

[30] *Supra*, ch. 2, at n. 47.

[31] See Rupp von Brünneck, 'Admonitory Functions of Courts: West Germany', 20 *Am. J. Comp. L.* 387 (1972); Linde, 'Admonitory Functions of Courts: United States', 20 *Am. J. Comp. L.* 415 (1972). The same technique has been repeatedly employed in the US in abrogating common law rules (*supra*, ch. 2, at n. 47), rarely in striking down a statutory provision on constitutional grounds. An example is *Shavers* v. *Kelley*, 402 Mich. 554, 267 NW 2d 72 (1978) where the Michigan court postponed the effect for 18 months to allow the Legislature (and the Commissioner of Insurance) to remedy the Act's deficiencies.

abstention or declaring a statute void for vagueness.[32] Both are devices to promote a colloquy between courts and legislature. It has been suggested that a similar technique should be developed for dealing with statutory desuetude, such as was conceivably plaguing guest statutes.[33] A prominent example of initiating such legislative review will be considered later.

That constitutional nullification would not promote that objective and should in any event be employed only with the greatest of reserve appears to have been realized in later cases. Thus, still in the context of guest passengers, a provision of the California Insurance Code permitting the exclusion of coverage for members of the insured's family in motor-car liability policies was upheld despite the discrimination *vis-à-vis* other passengers, the disappearance of the family immunity doctrine, and the previous dismissal of concern about fraud and collusion—on the realistic ground this time that the purpose of the exemption, well within legislative competence, was to encourage insurance by lowering premiums for car owners who either could not afford to pay more or had conceivably made other arrangements for their family's medical coverage.[34] This doctrinal reversal was reflected as much in what the court said as in what it did:

> That we may disagree with some or all of these reasons affords no justification whatever for the substitution of our own view of what is proper public policy for that of the Legislature ... [35] To hold otherwise and, in effect, to require family member liability coverage against the better judgment of the contracting parties would constitute an unprecedented judicial interference into private contractual and economic arrangements ...[36]

[32] See Bickel, *The Least Dangerous Branch: The Supreme Court at the Bar of Politics* 111-98 (1962).

[33] Calabresi, *A Common Law for the Age of Statutes* 10-11 (1982). *Weinrot* v. *Jackson*, 40cal. 3d 327, 708 p. 2d 682 (1985) de-construed an obsolete statute by confining an employer's statutory *per quod* action to domestic employment (cf. *I.R.C.* v. *Hambrook* [1956] 2d. B. 641).

[34] *Farmers Insurance Exchange* v. *Cocking*, 29 Cal. 3d 383, 628 P. 2d 1 (1981). That *Brown* v. *Merlo* dealt with a restriction on tort liability, not insurance, and that it was mandatory, not permissive, cannot account for the difference in approach or outcome. Significantly the court's opinion was delivered by Richardson, J., the only conservative appointee.

[35] Ibid., at 388-9. [36] Ibid., at 390-1.

1.3   Medical malpractice laws

Constitutional challenges have not been confined to statutes,
like the guest statutes, which had arguably become obsolete
and whose elimination could perhaps be explained as a mere
housekeeping or cleaning-up operation. More 'touchy' are
judicial confrontations with contemporary statutes whose
object is to reform the common law, even to correct recent
decisions of the court itself. Two major waves of statutory tort
reform, one dealing with medical, the other with automobile
liability, have provided a testing ground for this kind of
constitutional adjudication.

In the throes of the 'medical insurance crisis' of the
mid-1970s, statutes in many states sought to return medical
insurance to a more affordable level and induce insurers to
continue providing malpractice cover by means of changing
the underlying liability rules and conditions. These included
'caps' on damages, abolition of the collateral source rule,
awards of periodical payments in place of lump sums, compulsory arbitration, even restrictions on contingent fees. Since
several of these reforms modified or impaired the common law
right to damages for negligence, courts were peculiarly sensitive
to appeals for constitutional protection. In the result they
went different ways, some upholding the legislation on the
basis of the traditional federal rationality test, others nullifying
it on the basis of stricter state standards of equal protection
or by resort to 'due process' or special state constitutional
provisions.[37]

A small minority went to the length of classifying the right
to damages for personal injuries as a 'fundamental' right, like
the right of privacy, subject to strict scrutiny and inviolable
in the absence of a 'compelling' state interest.[38] A more
moderate and more widely favoured approach was to treat
the right to damages in tort, while not a fundamental right,
yet as so important that it requires restrictions to be subjected

---

[37] See Smith, 'Battling a Receding Tort Frontier: Constitutional Attacks
on Medical Malpractice Laws', 38 *Okla. L. Rev.* 195 (1985); Redish,
'Legislative Responses to the Medical Malpractice Insurance Crisis: Constitutional Implications', 55 *Tex. L. Rev.* 759 (1977).

[38] *White* v. *State*, 661 P. 2d 1272 (Mont. 1983).

to a more rigorous scrutiny than allowed under the rational basis test. This intermediate test, best expressed by the New Hampshire Court, was 'whether the challenged classifications are reasonable and have a fair and substantial relation to the object of the legislation. Whether a malpractice statute can be justified as a reasonable measure in furtherance of the public interest depends on whether the restriction of private rights sought to be imposed is not so serious that it outweighs the benefits sought to be conferred upon the general public.'[39] By this test, limitations on damages against health care providers have been found unjustifiable because it was considered arbitrary and unfair to single out the most severely injured malpractice victims for special relief to medical tortfeasors and their insurers. It might have been different if the burden had been spread among all injured, for example by reducing damages pro rata rather than by an absolute limit.

In the light of its chastening guest statute experience, the California Court's ruling was as eagerly awaited as it was surprising. A bare majority upheld the medical reform statute against several separate challenges of its principal provisions,[40] by returning as promised to the deferential standard of rationality.[41] All of these provisions trenched on the special interests of the trial bar, but especially the restrictions on contingent fees. The decision to uphold even those restrictions is all the more remarkable because they were less obviously related to the avowed legislative purpose of reducing insurance costs. Contingent fees come out of the plaintiff's award, and are not an added cost like the defendant's legal fees (which were not, however, touched). Besides, did the provision not discriminate against victims of medical negligence, to whom

---

[39] Carson v. Maurer, 120 NH 925, 424 A. 2d 825, 831, 12 ALR 4th 1 (1980).

[40] American Bank & Trust Co. v. Community Hospital, 36 Cal. 3d 359, 683 P. 2d 670 (1984) rev'g itself from 660 P. 2d 829 (1983) (periodical payments); Barme v. Wood, 37 Cal. 3d 174, 689 P. 2d 446 (1984) (collateral benefits); Roa v. Lodi Medical Group, 37 Cal. 3d 920, 695 P. 2d 1058, appeal dism. 106 S. Ct. 421 (1985) (contingent fees); Fein v. Permanente Medical Group, 38 Cal. 3d 137, 695 P. 2d 665, appeal dism. 106 S. Ct. 214 (1985) (collateral benefits, 'caps' on non-pecuniary damages).

[41] The minority adhered to the Merlo/Carson standard, despite its repudiation in Schwalbe v. Jones, supra, n. 26.

it alone applied, with the result that they would find it more difficult to procure the best legal representation? Yet these arguments did not persuade a majority of the Court who thought that the limitation on contingent fees did contribute to reduced insurance costs, because a plaintiff would be more likely to agree to a lower settlement, because it would deter attorneys from instituting frivolous suits or encouraging their clients to hold out for unrealistically high settlements, and because it would compensate plaintiffs for the other limitations on recovery, including the limitation for non-economic loss to $250,000.[42]

An even more dampening effect on litigation was intended by the Florida statute which provided for the award of reasonable attorney fees to the prevailing party (whether plaintiff or defendant).[43] This provision, unlike some others, in the medical liability statute was also upheld against constitutional challenge on the summary ground that forty other Florida statutes contained similar provision.[44] It was, however, eventually replaced by a statute limiting contingent fees on the more conventional model.[45]

1.4   Due process

Another constitutional guarantee to which opponents of tort statutes sometimes appeal for nullification is the prohibition in federal and state constitutions against depriving 'any person of life, liberty and property without due process of law'.[46]

[42] An empirical analysis found that fee ceilings reduced average settlement size by 9%, increased abandoned claims from 43% to 48%, and reduced litigated cases from 6.1% to 4.6%: Danzon and Lillard, *The Resolution of Medical Malpractice Claims: Modeling and Analysis* (1982).

[43] That this measure was inept to contain costs is illustrated by *Good Samaritan Hosp. Ass'n* v. *Saylor*, 495 So. 2d 782 (1986), awarding $1.1m. according to the federal lodestar formula, 2,000 hours of work at $275 per hour multiplied by a contingency factor of two. On fees see *infra*, ch. 6.

[44] *Florida Patient's Compensation Fund* v. *Rowe*, 472 So. 2d 1145 (1985).

[45] Fla. Stat. § 768.595 (1985).

[46] US Constitution, 5th and 14th Amendments. A cause of action is a species of 'property' protected by the due process clause, but the question is what process is due. Being inchoate until reduced to final judgment, it is a right very different from traditional ownership of tangible property. The focus of protection is therefore primarily on procedural fairness.

Whereas equal protection seeks to assure equal treatment between classes of individuals, due process demands fairness between the state and the individual. The US Supreme Court applies approximately the same two-tier standard as under equal protection: restrictions on fundamental rights or the political process are subject to strict scrutiny and can be justified only by a compelling countervailing interest. On the other hand, economic and social regulations are presumed valid unless arbitrary or capricious.[47] Particularly in view of the bitter controversy over the court's partisan use of the due process clause in opposition to social reforms in the earlier decades of this century during 'the *Lochner* era',[48] judicial restraint has since become particularly pronounced in dealing with denials of 'substantive' even more than with 'procedural' due process.

Substantive due process has been invoked against limitations on the right to damages for personal injury under medical liability as well as automobile no-fault statutes. Whether the total abolition of such a tort claim without substitute is constitutionally permissible is perhaps still an open question, but the US Supreme Court has noted that

Our cases have clearly established that a person has no property, no vested interest, in any rule of the common law. The Constitution does not forbid the creation of new rights, or the abolition of old ones recognized by the common law to attain a permissible legislative objective, despite the fact that otherwise settled expectations may be upset thereby. Indeed, statutes limiting liability are relative commonplace and have consistently been enforced by the courts.[49]

[47] See e.g. *Usery* v. *Turner Elkhorn Mining Co.*, 428 US 1 (1976).

[48] Associated with the notorious decision in *Lochner* v. *New York*, 198 US 45 (1905) and the 'Nine Old Men' who thwarted President Roosevelt's 'New Deal'. See Pearson and Allen, *The Nine Old Men* (1937); Jackson, *The Struggle for Judicial Supremacy* (1949); Tribe, *American Constitutional Law* 421–455 (1978).

[49] *Duke Power Co.* v. *Carolina Environmental Study Group*, 438 US 59, 88 n. 32 (1978). The Court upheld the Price–Anderson Act, limiting liability for nuclear accidents to $560m. but establishing an insurance syndicate and imposing strict liability. But the 37 states with so-called 'open court' guarantees make abolition, as distinct from regulation, of common-law causes of action virtually impossible. e.g., *Boswell* v. *Phoenix Newspapers, Inc.*, 730 P. 2d 186 (Ariz. 1986), striking down the Ariz. retraction statute.

One could add that otherwise all statutory reform of the common law might be stultified and courts would have the last word on all common law development.

A few causes of action which had become obsolete with changing cultural perspectives, like seduction and alienation of affections, have been widely abolished by statutes that have escaped constitutional challenge.[50] But the abolition of even viable causes of action has survived constitutional attack in several instances, among them the abrogation of non-pecuniary damages for retracted libels[51] and of all liability to the victims of drunk drivers negligently supplied by defendants with alcohol.[52]

Most of the medical liability statutes contain limitations on damages which, as previously noted, were defended with varying success against the charge of unjustifiably dis-criminating against, and thereby denying equal protection to, plaintiffs in malpractice actions. The California Court, by a precarious majority, upheld the limitation on non-pecuniary damages to $250,000 as also conformable to the requirement of due process, noting that 'thoughtful jurists and legal scholars have for some time questioned the wisdom of damages for pain and suffering in negligence actions', that the legislature was within its rights to conclude that it was in the public interest to obtain some cost savings by this device, and that, in any event, it knew of 'no principle of California—or federal—constitutional law which prohibits the Legislature from limiting the recovery of damages in a particular setting in order to further a legitimate state interest'.[53] In support it relied on the US Supreme Court's upholding the limit in a federal statute on liability in the event of a nuclear accident.[54] And it distinguished contrary decisions on the ground that they dealt with statutes imposing an arbitrary limit on all

[50] See Prosser and Keeton, *supra*, at 929–30.

[51] *Werner* v. *So. California Newspapers*, 35 Cal. 2d 121, 216 P. 2d 825 (1950).

[52] *Cory* v. *Shierloh*, 29 Cal. 3d 430, 629 P. 2d 8 (1981).

[53] *Fein* v. *Permanente Medical Group*, *supra*, at 161. The US Supreme Court dismissed an appeal: 106 S. Ct. 214. Only White, J., dissenting, questioned 'whether due process required a legislatively enacted compensation scheme or a *quid pro quo* for the common law or state law remedy it replaces, and if so, how adequate it must be': at 216.

[54] *Duke Power Co.* v. *Carolina Environmental Study Group*, *supra*.

damages instead of merely on non-pecuniary damages. But does this distinction not concede that a statute may not deprive a plaintiff of adequate recovery, either in the amount or the kind of damages targeted? Other courts have also questioned whether there was an insurance crisis and whether, even if the original legislative wisdom as to the chosen means in combating it must be respected, time may not have shown that their perpetuation now constitutes a denial of due process.

Where the common-law cause of action is replaced by a compensation scheme, the long fight over the validity of workers' compensation in the first twenty years of this century eventually ended in acceptance of the notion of a 'trade-off' as satisfying minimal requirements for due process.[55] By analogy, automobile no-fault laws imposing 'thresholds' for tort claims have in general been successfully defended on the ground that the loss was balanced by an 'adequate substitute' in the shape of no-fault benefits.[56] On the other hand, the Florida Court nullified the total abolition of tort actions for damage to motor vehicles under $500 because it lacked any reasonable substitute, such even as mandatory insurance.[57] This conclusion is difficult to justify because voluntary first-party insurance remained available to any car owner prepared to pay for it; the statutory design simply chose to give an option to risk-preferrers instead of paternalistically treating all car owners willy-nilly as risk-avoiders.

By contrast, the Michigan Court drew a sensitive distinction between property damage and personal injury. It upheld the mandatory property insurance provision which covers only damage caused by the insured to the property of third parties

[55] *New York Central Railroad Co.* v. *White*, 243 US 188 (1917).

[56] See King, 'Constitutionality of No-Fault Jurisprudence', 1982 *Utah L. Rev.* 797 (state-by-state analysis); Siedel, 'The Constitutionality of No-Fault Insurance: The Courts Speak', 26 *Drake L. Rev.* 794 (1977).

[57] *Kluger* v. *White*, 281 So. 2d 1 (Fla. 1973). See Vinson, 'Constitutional Stumbling Blocks to Legislative Reform', 15 *Fla. St. U. L. Rev.* (1987). So also many courts have struck down 'periods of repose' under products liability statutes, for violating due process or 'access to courts' provisions, on the ground that the abrogation or limitation of an existing legal remedy can be justified only if it is a reasonable and not arbitrary measure for eliminating a clear social or economic evil: see *Berry* v. *Beech Aircraft Corp.*, 717 P. 2d 670 (Utah 1985).

other than their cars, as justified by anticipated improvements of the accident reparation system in general, in deference to legislative judgment in matters of economic control. On the other hand, changes relating to personal injury demanded closer scrutiny in order to ensure fairness in the determination of rates and policy terms and effective control of insurers by administrative agencies.[58]

'Procedural' due process has also occasionally proved an impediment to tort reform. Although stemming from the same constitutional clause, this requirement is applied more stringently for a number of reasons, the most important of which are the greater confidence courts have in evaluating procedural fairness and the less intrusive nature of their decisions in regard to substantive policy. A strong illustration is the US Supreme Court's decision in *Bell* v. *Burson*,[59] disallowing a typical provision of a state financial responsibility law under which a driver's licence was suspended after involvement in an accident without a hearing on the question of liability, unless he could furnish security for the amount of damages claimed by an aggrieved party. Fault and liability were not, as argued, irrelevant to the scheme, when viewed for substance rather than form. For example, suspension was lifted in case of a release or adjudication of no liability. Thus the court ruled that a state could not, consonant with due process, eliminate consideration of liability in its prior hearing, because the suspension impinged on an important interest of the licensee.[60] In the result, financial responsibility laws had to be recast or replaced by compulsory insurance.[61]

Automobile no-fault and medical liability statutes have mostly had to contend with equal protection and substantive due process. However, both have been attacked also for abrogating jury trial, generally under the specific constitutional guarantee

[58] *Shavers* v. *Kelley*, 402 Mich. 554, 267 NW 2d 72 (1978).

[59] 402 US 535 (1971).

[60] 'A procedural rule that may satisfy due process in one context may not necessarily satisfy procedural due process in every case. Thus procedures adequate to determine a welfare claim may not suffice to try a felony charge. Clearly, however, the inquiry into fault and liability requisite to afford the licensee due process need not take the form of a full adjudication of the question of liability': ibid. at 540.

[61] *Infra*, ch. 5, at 83.

under the federal and state constitutions. These challenges have mostly failed, in the automobile case on the ground that the statute did not replace the jury as trier of fact but rather modified the substantive law.[62]

Pre-trial screening provisions under several medical liability statutes have been tested with varying results. Some have been held to infringe separation of powers or access to courts guarantees,[63] as was a mandatory provision for arbitration in lieu of jury trial in one automobile statute regarding property damage under $3,000.[64] Most screening provisions, however, have been upheld as not violating the right to a fair and impartial jury.[65]

2. NON-CONSTITUTIONAL NULLIFICATION

That judicial distaste for a particular statute not infrequently leads to contrived and cavalier constructions to narrow its range of application is a commonly observed technique in American courts.[66] If this abuse of self-proclaimed judicial neutrality was once castigated as the monopoly of conservative judges dealing with liberal reforms, it has in more recent times been practised with equal zeal by liberal judges confronted with statutes embodying (to them) repugnant ideas or methods.

---

[62] e.g. *Montgomery* v. *Daniels*, 38 NY 2d 41, 340 NE 2d 444 (1975).

[63] e.g. *Bernier* v. *Burris*, 113 Ill. 2d 219, 497 NE 2d 763 (1986) (but upholding other provisions).

[64] *Opinion of the Justices*, 113 NH 205, 304 A. 2d 881 (1973); *Grace* v. *Howlett*, 51 Ill. 2d 478, 283 NE 2d 474 (1972).

[65] See Smith, *supra*, at 213-14.

[66] e.g., in *Ault* v. *International Harvester Co.*, 13 Cal. 3d 113, 528 P. 2d 1148 (1978) a statute which made evidence of subsequent repairs inadmissible 'to prove negligence or *culpable* conduct' was held inapplicable to a claim for strict liability; in *Nelms* v. *Laird*, 406 US 797 (1972) strict liability did not qualify as a 'negligent or *wrongful* act or omission'. (Ital. added.) By the same token, statutes are sometimes distorted in the opposite direction in order to support a desired conclusion. A particularly crass example is *Royal Globe Insurance Co.* v. *Superior Court*, 23 Cal. 3d 880, 592 P. 2d 329 (1979) where the California Court deliberately misconstrued a regulatory penal statute, aimed against certain 'practices' by an insurer, into providing also a private remedy for a tort victim for a single bad faith refusal to settle. See *infra*, ch. 5, at n. 141.

A revealing illustration was the technique employed by the California Court to overcome a statutory hurdle in its desire to replace the complete defence of contributory negligence by comparative fault.[67] The obstacle was Section 1714 of California's Civil Code, enacted in 1872, which proclaimed magisterially that

everyone is responsible not only for the result of his willful acts, but also for an injury occasioned to another by his want of ordinary care or skill in the management of his property or person, *except so far as the latter has, willfully or by want of ordinary care, brought the injury upon himself.*

The court first rejected the plaintiffs' argument that 'except so far as . . .' had all the time been intended to introduce a regime of comparative negligence, even if no one had hitherto had the perspicacity to notice it. The historical evidence was simply too strong against any intention by the celebrated codifier, David Dudley Field,[68] let alone the California Law Commissioners, to introduce such a drastic and precocious reform so surreptitiously. Instead, the court advanced the surprising rationale that 'the code's peculiar character as a continuation of the Common law permit[s] if not require[s] that section 1714 be interpreted so as to give dynamic expression to the fundamental precept which it summarizes'.[69] Just as the Code section implicitly acknowledged the 'last clear chance' exception, it also countenanced judicial modifications of the defence of contributory negligence in the future. Thus, as Professor England has pointed out,[70] the court in one swoop renegued both on the common law principle of statutory preclusion and on the civil law tradition of Code interpretation, in order to justify its role of law reformer.[71]

[67] *Li* v. *Yellow Cab Co.*, 13 Cal. 3d 804, 532 P. 2d 1226 (1975).

[68] On the code movement in the United States see Cook, *The American Codification Movement* (1981); Friedman, *A History of American Law* 351–5 (1973); also Van Alstyne, 'The California Civil Code', *West's Annot. Calif. Codes, Civil Code* 1.

[69] *Supra*, n. 64 at 822. Once acknowledged that a common law rule can be judicially changed, why should a statutory rule that is a mere continuation thereof (Civil Code art. 5) be any more sacrosanct?

[70] England, *Li* v. *Yellow Cab Co.*: 'A Belated and Inglorious Centennial of the California Civil Code', 65 *Calif. L. Rev.* 4 (1977).

[71] Cf. a similar manoeuvre in *Vincent* v. *Pabst Brewing Co.*, 47 Wis. 2d

Later, the same court resorted to even less conventional methods of nullifying statutes that stood in the way of a 'better' rule.

2.1 Contribution statutes

Contribution among tortfeasors was sponsored by the Uniform Law Commissioners in two successive model acts in 1939 and 1955. Both embodied the principle of contribution pro rata, i.e. apportionment not in accordance with the tortfeasors' respective shares of fault or responsibility but equally. This had the advantage of facilitating settlements and shortening trials by dispensing with the need to compare the defendants' demerits. Besides, it may have been thought that the controlling principle was causation and that comparison of causes was even more mind-boggling than comparison of fault.[72]

An additional feature of the legislation, as adopted in several states including New York, Michigan, and California, was to allow contribution only among tortfeasors who had been sued to a joint judgment. This restriction had the admitted disadvantage of perpetuating the plaintiff's power, as at common law, to control at his whim whether the loss was to be shared or borne only by those defendants he selected. More important, however, seemed to be avoidance of inconsistent awards, disputes over the amount of the plaintiff's loss, what effect to attach to a prior settlement, lapse of time, and so forth. It also promoted administrative efficiency by deterring multiple litigation. Admittedly, those objectives could also be attained by at least permitting cross-complaints, but only Michigan allowed this modification to satisfy the joint judgment requirement.[73] It was opposed by plaintiffs' attorneys who did not wish to have foisted on them defendants who

120, 177 N, W. 2d 513 (1970) where three judges out of seven accepted the argument that a statute authorizing apportionment for plaintiffs less than 51% at fault, did not preclude apportionment for more culpable plaintiffs, for whom the statute had left intact the common law rule, which could accordingly be judicially modified.

[72] Cf. 'causation itself is difficult enough; degrees of causation would really be a nightmare': Chapman, 64 *LQR* 26, 27 (1948).

[73] Repealed in 1974 on abandoning the principal rule: *Mich. Comp. Laws* Ann. § 600.2925.

might evoke special sympathy, leading to lower verdicts. Moreover, the usual reason for not joining a particular co-tortfeasor is that he is uninsured or a relative or friend of the plaintiff whom the latter wished to spare.[74]

Both these features of the New York statute came into question in *Dole* v. *Dow Chemical Co.*,[75] later described as 'the celebrated decision' by the California Court which adopted it.[76] The deceased worker had been overcome by fumes from a chemical he was using when cleaning the inside of a big tank. His family made claim against Dow Chemical Co. for failing to issue warnings concerning the use of the cleanser in a confined space; Dow, in turn, claimed contribution from the deceased's employer for negligence in not testing the chemical before use and warning its employees. This cross-claim was impeded by two statutory obstacles: first, New York's Workers' Compensation Act, like that of all other American jurisdictions, makes it the exclusive remedy against the employer;[77] secondly, the Contribution Act, as already noted, permitted contribution only among tortfeasors held liable in a joint judgment. Thus, there was no common liability in tort to the deceased or his dependants, still less any joint judgment.

None of this, however, fazed the Court of Appeals, New York's highest court. The contribution statute, it ventured to think, was simply a partial legislative modification of the harsh common law 'no contribution' rule, which did not preclude a parallel judicial extension of common law 'indemnity'. True, in the past 'indemnity' had meant a shifting of the complete loss,[78] but 'contribution' need not pre-empt all sharing of the loss. Henceforth, there could also be 'partial indemnity', responsive to the precept of equitable apportionment. This would dispense with three statutory restrictions: joint tort liability, joint judgment, and pro rata shares. Thus by merely

---

[74] As was indeed the case in *American Motorcycle Assn.* v. *Superior Court*, 20 Cal. 3d 578, 578 P. 2d 899 (1978).

[75] 30 NY 2d 143, 282 NE 2d 288 (1972). In 1974 the Contribution Act (CPLR §§ 1401-2) was amended, abolishing the 'joint judgment' requirement and introducing 'equitable shar[ing]'.

[76] *American Motorcycle Assn.* v. *Superior Court, supra,* at 597.

[77] See Larson, *Workers' Compensation,* vol. xii.

[78] Cf. *Fairfield* v. *McGrath,* (1984) 2 NSWLR 247, 249 (Glass, J A): 'only a 100% contribution [can] be called an indemnity.'

relabelling contribution 'partial indemnity', three statutory provisions were in effect set aside. The California Court eagerly followed the same precept.[79]

The cynicism with which a legislative intent is conjured to support a desired judicial decision, despite the unequivocal wording of the statute to the contrary, found further illustration in the cavalier manner of either adopting or rejecting the statutory model for subsidiary questions. One such question was whether to retain the traditional 'joint and several liability' rule, under which each co-defendant is liable to the plaintiff for the full amount of the judgment regardless of his own share of responsibility. This rule in practice often leaves the defendant with the deepest pocket to bear the whole or a disproportionately large share of the judgment. Should it not now be abolished in deference to the over-arching mandate that liability should not exceed one's share of fault—a self-imposed mandate that had inspired the court to introduce comparative negligence and comparative contribution? On this issue, however, the California Court preferred constancy in the cause of plaintiffs to constancy of principle by upholding the 'joint and several liability' rule even when the plaintiff was himself at fault.[80] So also it did in its decision to adhere to the rule of the Contribution Act that a non-settling defendant can claim credit only for the amount of a prior (good faith) settlement, not for the share of the settlor's fault, and to the rule that the latter is thereafter free from any claim for contribution.[81]

[79] *Supra*, n. 74. See Fleming, 'Report to the Joint Committee of the California Legislature on Tort Liability on the Problems Associated with *American Motorcycle Assn.* v. *Superior Court*', 30 *Hast. L. J.* 1465 (1979).

[80] *American Motorcycle Assn.* v. *Superior Court, supra*. The lower appellate court had found 'several' liability the better way to reconcile the pro rata rule for contribution under the Contribution Statute with the *Li* mandate. The Supreme Court resolved the impasse by side-stepping the pro rata rule and retaining 'joint and several' liability. Elsewhere, of the 9 states that follow the rule of non-contribution, the comparative negligence statutes of 4 (Ind., Kan., NH, Vt.) have eliminated joint and several liability. N.Mex. found that the judicial adoption of comparative negligence required 'several' liability notwithstanding the Uniform Act; Okla. opted for 'several' liability only when the plaintiff is also negligent.

[81] In *Baget* v. *Shepard*, 180 Cal. Rptr. 396 (1982) an intermediate appeal court departed from this model, reducing the non-settling defendant's li-

In contrast, on the question whether, in cases of dual fault and dual loss, the parties' claim and counterclaim should be set-off, the California Court reverted to the technique of explaining away an obstructive statute in order to help plaintiffs and their lawyers to a larger recovery.[82] Set-off, as mandated by statute,[83] being an equitable principle, would lead to self-defeating inequitable results if applied without reference to the parties' insurance coverage. Accordingly, the court proclaimed, the statute 'cannot properly be interpreted to require set-off in cases in which such a set-off will defeat the principal purpose of the financial responsibility law and will provide an equitable windfall to an insurance carrier at the expense of the carrier's insured'.[84]

This episode would not be complete without a postscript. The 'joint and several liability' rule, so staunchly defended by the California Court, later became the target of a reform campaign spearheaded by local governments in alliance with the insurance industry and other 'deep pocket' groups. After efforts to procure reform by legislation were foiled by the plaintiffs' bar and its legislative supporters, a popular initiative eventually approved a measure (dubbed the 'deep pocket' proposition) to eliminate the 'joint and several liability' rule with respect at least to claims for non-economic loss.[85] What turned the scales in this instance was that popular opposition to the rule could be enlisted because of its immediate fiscal impact on taxpayers as a result of its application against local government bodies for road hazards which may have contributed only a minor share in respect of accidents by un- or under-insured[86] drivers. Five months later, three justices

ability by the released tortfeasor's proportionate share. The opinion was decertified. See Grodin, 'The Depublication Practice of the California Supreme Court', 72 *Calif. L. Rev.* 514 (1984). But much to the consternation of the plaintiffs' bar, *Tech–Bilt* v. *Woodward–Clyde*, 38 Cal. 3d 488, 698 P. 2d 159 (1985) recently reversed a prior trend by holding that good faith called for proportionality: the settlement must be fair to non-settling defendants.

[82] *Jess* v. *Herrmann*, 26 Cal. 3d 131, 604 P. 2d 208 (1979).

[83] Code of Civil Procedure §§ 431.70 and 666.

[84] *Jess* v. *Herrmann, supra*, at 143.

[85] Fair Responsibility Act of 1986, amending Civil Code §1431. Some states, e.g. Arizona, abolished the rule entirely.

[86] The rule had become increasingly burdensome as the result of two

were defeated at the polls. If this denouement has any moral, it is that judges who abandon the tradition of open-mindedness and neutrality in pursuit of their own political agenda risk being treated for what they are—politicians—and, in a democracy, liable to the sanctions of popular displeasure.

3. MORE CONSTITUTIONALISM: FREE SPEECH

The preceding two sections laid emphasis primarily on the judicial evisceration of *statutes*, a process almost exclusively carried forward by state courts. The principal *common law* casualties of constitutional doctrine have been the twin torts of defamation and invasion of privacy, at the hands of federal courts.[87]

3.1. Defamation

It is a truism that the law of defamation in a modern democratic society is dominated by the need to balance its solicitude for individual reputation against the public's interest in freedom of speech and access to information and opinion, especially on matters of public concern. The complexity of the contemporary English law not only harbours surviving relics of an earlier political culture but reflects continuous efforts over more than a century to readjust that balance by patchwork legislation and modest judicial innovations. But in comparison with American values, it reveals the much greater concern of English public opinion to safeguard the reputation and privacy of the individual, due in part to a deeper scepticism as to whether uninhibited free speech is indeed the least costly way of disseminating worthwhile information, in part to a much more ambivalent attitude towards the blessings

developments: first, the dramatic increase in the size of awards and exposure not matched by corresponding increases in insurance cover of motorists who are therefore frequently underinsured; secondly, the tendency to tap deep pockets among those whose responsibility is more remote and tenuous.

[87] Civil rights actions, the so-called 'constitutional torts', fall outside the present context, being complementary to, rather than modifying, traditional common law doctrine.

of the media. The contrast between the two legal cultures is particularly striking in regard to political speech. The concessions afforded by English law to the public's 'right to know' about matters of public interest are extremely circumspect and leave a good deal of protection for politicians and other public figures against investigative reporting. American law, on the other hand, reflects 'a profound national commitment to the principle that debate on public issues should be uninhibited, robust and wide open'.[88]

The English libel law accordingly underwent several common-law modifications in its trans-Atlantic environment. Its most controversial feature was, and remains, its commitment to strict liability, placing the risk of error, even faultless error, squarely on the media and other defendants. This stringent liability is mitigated only by narrow defences: besides justification, by so-called privileged occasions and fair comment. Significantly, there is no privilege to communicate, even in good faith, defamatory statements to the public at large on matters of public interest, nor is fair comment on a matter of public interest a defence unless it is 'fair' (e.g. not inspired by ill will) and based on true facts.

The strict liability rule was widely modified in many American states by treating libels not defamatory on their face (libels *per quod*) like slander (oral defamation), i.e. not actionable in the absence of material loss.[89] Thus it is only when the impugned expression carries its own warning of potential trouble that the writer is placed at risk, essentially for fault in failing to pursue inquiries. Another modification, anticipating future developments, was that, in a number of states, the defence of comment was applied also to inaccurate facts.[90] Both of these state-sponsored modifications favouring the media have since become overshadowed by the federal intervention in 1964 which created a vast new sanctuary for

---

[88] *New York Times* v. *Sullivan*, 376 US 254, 270 (1964). For valuable comparisons with English and German law see Barendt, *Freedom of Speech* (1987).

[89] Harper, James, and Gray, *supra*, § 5.9A. This bears comparison with the belated introduction in 1954 of the statutory English defence of 'innocent publication', subject to an offer of amends (Defamation Act 1956 s.4).

[90] e.g. *Coleman* v. *MacLennan*, 78 Kan. 711, 98 P. 281 (1908).

defamation in the name of constitutionally protected free speech.[91]

Until the seminal decision in *New York Times* v. *Sullivan*[92] it was the accepted belief that defamation, no more than obscenity, could enlist the First Amendment guarantee of 'freedom of speech or of the press'.[93] This innocent assumption became a casualty of the civil rights struggle in the South. The case concerned a libel action by the Police Commissioner of Montgomery, Alabama, for an advertisement in the *New York Times* which reported, with minor factual inaccuracies, strong-arm 'police action' against civil rights demonstrators. A bigoted jury, inflamed against the Eastern press for its support of racial integration, returned a verdict for $500,000, which was sustained by the Alabama Supreme Court as conformable with state law, including a jury instruction equating malice with ill will and despite the requirement that the libel must sufficiently identify the plaintiff. The US Supreme Court responded by ruling that an elected public official could recover for defamation only on proof of 'actual malice', i.e. knowing that the impugned allegation was false or made with reckless disregard for the truth. In a case such as this involving political speech, the core area of the First Amendment, the constitutional concern for free speech required no lesser measure of subordination for individual reputation of public officials. Erroneous statements are inevitable in fearless and robust debate and must be tolerated in order to assure it necessary breathing-space; only calculated falsehood was devoid of social value. The sacrifice thus demanded from public officials was in any event mitigated by their ample opportunities for public rebuttal and by the absolute immunity they themselves enjoyed for any statements concerning their official duties.[94] While bearing some re-

[91] That a similar adjustment may be required by the European Convention on Human Rights, Art. 10 (1) ('everyone has the right to freedom of expression') is foreshadowed by the Report of the Commission in *Lingens* v. *Austria* (1984). See Elder, 'Freedom of Expression and the Law of Defamation: The American Approach to Problems Raised by the *Lingens* Case', 35 *Int. & Comp. L. Q.* 891 (1986).

[92] *Supra*, n. 88.

[93] See *Chaplinsky* v. *New Hampshire*, 315 US 568, 571-2 (1942).

[94] *Barr* v. *Matteo*, 360 US 564 (1959).

semblance to common-law qualified privilege, the constitutional defence is not forfeited merely by any hostile or
other improper motive. Not only is such motive a common
accompaniment of political speech, but it would have been
open to abuse by biased juries. For that reason also, proof of
'actual malice' had to meet the constitutional standard of
'clear and convincing evidence', reviewable *de novo* by appellate
courts.

Ever since, the US Supreme Court has been embroiled in
exploring the implications of constitutionalizing the state law
of defamation. It first extended the ruling to 'public figures'[95]
and, at its furthest reach, to all 'matters of public concern'.[96]
Thereafter increasing doubts both about the extent of justified
constitutional intrusion and the appropriate formula signalled
retreat. In the landmark case of *Gertz* v. *Welch*[97] a narrow
majority repudiated the extension of the *New York Times*
requirement of actual malice beyond public figures: only from
the latter could such a sacrifice be demanded as the price for
voluntarily seeking the public limelight.[98] But as a makeweight,
the court prescribed that henceforth liability for defamation
in general had to be based on 'fault' and that neither presumed
nor punitive damages could be awarded in the absence of
'actual malice'. This first-time adversion to the issue of damages
was a belated recognition of the fact that extravagant awards
against the media had really provided the impetus for
constitutional controls; it also called into question the choice
of status immunity ('public figures') rather than control of
damages as providing the best constitutional balance.

[95] *Curtis Publishing Co.* v. *Butts* and *Associated Press* v. *Walker*, 388 US 130
(1967).

[96] *Rosenbloom* v. *Metromedia*, 403 US 29 (1971). This proposition was
supported only by a 'plurality' of 4, the 5th member constituting the majority
basing himself on a different ground.

[97] 418 US 323 (1974). Justice Blackmun deferred his personal views to the
need of forming a majority.

[98] 'Public figures' for the most part are persons who had 'assumed roles of
a special prominence in the affairs of society. Some occupy positions of such
persuasive [*sic*] power and influence that they are deemed public figures for
all purposes. More commonly, [they] have thrust themselves to the forefront
of a particuar controversy in order to influence the resolution of issues involved' (ibid., at 345). The trend to take a narrow view has continued since
Gertz: *Time, Inc.* v. *Firestone*, 424 US 448 (1976).

Almost in passing, the *Gertz* Court also proclaimed that

[u]nder the First Amendment there is no such thing as a false idea. However pernicious an opinion may seem, we depend for its correction not on the conscience of judges and juries, but on the competition of other ideas. But there is no constitutional value in false statements of fact.[99]

This principle flows from the long constitutional tradition of a 'market place of ideas'[100] but it still awaits authoritative explication. The *Restatement, Second* interpreted it as conferring an absolute immunity on 'simple expression of opinion based on disclosed or assumed non-defamatory facts . . . no matter how unjustified and unreasonable the opinion may be or how derogatory it is'.[101] The US Supreme Court itself has only had occasion to apply it to 'rhetorical hyperbole', construed as non-defamatory in the particular context;[102] never to a defamatory deduction from, or evaluation of, disclosed or assumed facts. Moreover, it has since become doubtful whether the privilege will apply to private matters, instead of remaining restricted, as common law 'fair comment' was, to matters of public interest.[103] Notably the defence is not limited, as was its common law forerunner, by a requirement of 'fairness'; presumably because recipients are able to evaluate the worth of an opinion whatever its motivation. Not that the fact/opinion distinction is free from ambiguity, but it has so far been applied liberally, having regard both to the disputed statement's context and to verifiability.[104]

In the meantime, the court remains fractured between a conservative group seeking to roll back federal intrusion into

---

[99] *Gertz* v. *Welch, supra,* n. 97.

[100] *Abrams* v. *US,* 250 US 616, 630 (1919) (Holmes J. diss.).

[101] See *Restatement of Torts, Second* § 566.

[102] *Greenbelt Publ. Assn.* v. *Bresler,* 398 US 6, 14 (1970) ('blackmail'); *Letter Carriers* v. *Austin,* 418 US 264, 285-6 (1974) ('A scab is a traitor to his God, his country, his family and his class').

[103] See *infra,* n. 105.

[104] See especially *Ollman* v. *Evans,* 750 F. 2d 970 (en banc DC Cir. 1984); cert. denied, 471 US 1127 (1985) (professor is 'an outspoken exponent of Marxism') in which a variety of views were expressed. Rehnquist, J., joined by Burger, CJ, would restrict the defence to political 'ideas', not 'opinions' in general: ibid., at 2663-4. Generally, Note, 'Statements of Fact, Statements of Opinion, and the First Amendment', 74 *Calif. L. Rev.* 1001 (1986).

state law, and the liberal wing, committed to the widest reach of free speech. In its latest decision, the former rallied a majority behind freeing state law of defamation from the *Gertz* requirement with regard to speech of a 'purely private concern'.[105] Although it now seems to be recognized that the constitutional shield is not confined to media defendants,[106] its restriction to matters of public concern will in effect terminate the constitutionalization of 'private' defamation.

3.2   Privacy

American law has been much more eager than the English to promote civilized standards of life in a congested world by judicially developing a specific tort remedy for invasion of privacy. This divergence can be explained on a number of grounds. One is the general disinclination of English courts, in contrast to the American, to recognize new causes of action without legislative authorization. Another is a legislative preference in England for criminal sanctions, which has been exercised in several respects against abuses of privacy,[107] in contrast to the American judicial development of a tort cause of action sounding in damages. A third is that opposition by the media has naturally been more effective on the legislative than on the judicial scene.

The impetus for developing a distinct action for invasion of privacy was a celebrated law review article in 1890 by Warren and Brandeis, the latter destined for a distinguished career in the US Supreme Court.[108] The article built a powerful case for the explicit recognition of such a right on the basis of common-law analogies, mostly English, garnered from nuisance and trespass to defamation and copyright. From this seed a large body of case law has grown up, now categorized in

---

[105] *Dun and Bradstreet* v. *Greenmoss Builders*, 472 US 749 (1985). The majority found it unnecessary to comment on whether this included the fault requirement. Two justices among the majority favoured total reversal of *Gertz*, i.e. even on matters of public concern.

[106] Ibid., at 773, 781–4.

[107] See Seipp, 'English Judicial Recognition of a Right to Privacy', 3 *Oxf. J. Leg. St.* 325 (1983).

[108] 'The Right to Privacy', 4 *Harv. L. Rev.* 193 (1890).

the *Second Restatement*.[109] Apart from intrusion into territorial privacy, an extended form of trespass, most types of invasion involve offensive publicity, especially public disclosure of private facts and commercialization of a person's name or likeness. In developing this tort, courts have on the whole been alive to the need for balancing the plaintiff's interest against the public's 'right to know'. This has been done by insisting on the intrusion being highly offensive by prevailing standards of decency and by recognizing 'newsworthiness' as a defence.[110]

But it was not until after *New York Times* that defendants awoke to the possibility of also enlisting constitutional protection. Their argument is strongest in cases of true and non-defamatory speech on matters of public interest. Contemporaneous media publication of the names of rape victims and juvenile offenders, if included in court records, has accordingly been held constitutionally protected.[111] State courts have not been unanimous on the legitimacy of resuscitating events long past, such as the identification of one-time prostitutes or criminals (defamatory but true), which presents a serious conflict between the public interest in rehabilitation and the educational value of publicity.[112] Little less troublesome are cases involving false but non-defamatory publicity: 'putting the plaintiff in a false light'. On two occasions the US Supreme Court invoked the *New York Times* standard of 'actual malice', thus tolerating liability for a known falsehood, however well-meaning the motive.[113] But the court has yet to pronounce on fictionalized biography and works intended as pure fiction yet susceptible of being

---

[109] § 652A–I, divided into four categories: 'Intrusion from Seclusion', 'Appropriation of Name or Likeness', 'Publicity Given to Private Life', and 'Publicity Placing Person in False Light'.

[110] See Prosser, 'Privacy', 48 *Cal. L. Rev.* 383 (1960); Prosser and Keeton, *supra*, ch. 20; Harper, James, and Gray, *supra*, ch. 9.

[111] *Cox Broadcasting Corp.* v. *Cohn*, 420 US 469 (1975).

[112] See *Sidis* v. *F R Publishing Corp.*, 113 F. 2d 806 (1940); *Briscoe* v. *Reader's Digest Ass'n.*, 4 Cal. 3d 529, 483 P. 2d 34 (1971).

[113] *Time, Inc.* v. *Hill*, 385 US 374 (1967); *Cantrell* v. *Forest City Publishing Co.*, 419 US 245 (1974).

understood to refer to an existing person.[114] In either case, freedom of publicity is peculiarly affected.

The category of 'giving publicity to private facts' least implicates First Amendment concerns. At first blush, it might seem that the publication of true and non-defamatory facts deserves at least as much constitutional protection as the publication of false and defamatory speech. But, in truth, in competition with freedom of information, privacy has higher credentials than reputation, precisely because it concerns matters of private interest whereas reputation implicates a public concern.[115]

---

[114] See e.g. *Spahn* v. *Julian Messner, Inc.*, 21 NY 2d 124, 233 NE 2d 840 (1967); *Hicks* v. *Casablanca Records*, 464 F. Supp. 426 (SDNY 1978); *Guglielmi* v. *Spelling–Goldberg Productions*, 25 Cal. 3d 860, 870-1, 603 P. 2d 454 (1979).

[115] See Nimmer, 'The Right to Speak from *Times* to *Time*: First Amendment Theory Applied to Libel and Misapplied to Privacy', 56 *Calif. L. Rev.* 935 (1968); Hill, 'Defamation and Privacy Under the First Amendment', 76 *Colum. L. Rev.* 1205 (1976).

4

THE JURY

1. GENERAL CHARACTERISTICS

Jury trial is the hallmark of the American tort system. Unknown to the civil law, it has all but disappeared from other common law jurisdictions except in a few remnants of the old Commonwealth.[1] In the United States, however, the civil jury has retained its august position, entrenched both by the federal and by most state constitutions.[2] It is uniquely associated with tort litigation, both because it functions almost exclusively in tort cases[3] and because the right to jury trial is rarely waived. Its prevalence is indicated by the following statistics: California, with a population of 25m. recorded approximately 100,000 tort claims in 1984–5; almost 70,000 dispositions, 1,660 of which were contested. Among the last were 1,560 jury trials, constituting 2 per cent of total dis-

---

[1] See Fleming, *Law of Torts* 284 (7th edn. 1987); Cornish, *The Jury* ch. 8 (1968); Devlin, *Trial by Jury*, ch. 6 (1966). In England, civil juries can be had as of right only in cases of defamation, fraud, malicious prosecution, and false imprisonment: Supreme Court Act 1981 s. 69. There are not more than two dozen civil jury trials each year: Zander, *Cases and Materials on the English Legal System* 384 (4th edn. 1984).

[2] US Constitution, 7th Amendment. States: e.g. Cal. Constitution, art. 1, § 7. Preserved is the right to jury trial as it existed under English law at a particular time, either 1791 under the 7th Amendment or the date of the first State constitution, e.g. 1850 in California. There is much learning on the application of this historical test to new rights and remedies created thereafter, but common law tort actions are clearly included. See, e.g., James and Hazard, *Civil Procedure* § 8.1 (3rd edn. 1985).

[3] Peterson and Priest, *The Civil Jury: Trends in Trials and Verdicts, Cook County, Illinois, 1960–1979* 6 (1982) note that fewer than 2% of jury trials involved non-tort issues, such as anti-trust (federal), condemnation in state but not federal cases.

positions.[4] Civil jury trials throughout the nation number in excess of 30,000 annually, most of them presumably in tort cases.[5]

The traditional veneration of the jury in American folklore has deep roots. In colonial days juries acquired an aura as a bulwark of liberty, willing to follow their conscience rather than the dictates of an alien authority. Until the beginning of the nineteenth century juries exercised law as well as fact-finding functions with scant control by the judiciary.[6] Later, the democratic ideals of the Jacksonian era perpetuated the belief in the jury as the genuine voice of the people and fountain of justice. Thus the American tradition of jury independence is as old as the institution itself. For the most part, of course, the jury's role to capture public attention was, and continues to be, in relation to criminal trials rather than civil litigation.[7] Change came when the social and economic transformation in the latter part of the nineteenth century began to undermine the assumption that the man in the street would continue to share the values, and serve the goals, of the new élites.[8] As a result, judicial control was tightened and

---

[4] Judicial Council of California, *1986 Annual Report* 138-9. In FY 1974-75 jury trial amounted to 4.3% of dispositions.

[5] See *State Court Caseload Statistics: Annual Report 1984*, Table 19 (1986).

[6] See Nelson, *The Americanization of the Common Law* 26-30 (1975) for the early Massachusetts experience.

[7] The most salient aspect of unfettered jury discretion is their *power* of nullification in criminal cases, tracing back to *Bushell's case* (1670) 6 Howell's St. Tr. 999, which permits acquittal or conviction of lesser crime contrary to judicial instructions. Even today, some social activists claim a *right* of nullification in 'conscience cases'. See Simson, 'Jury Nullification in the American System: A Skeptical View', 54 *Tex. L. Rev.* 488 (1976); Scheflin and Van Dyke, 'Jury Nullification', 43:4 *Law & Contemp. Prob.* 51 (1980). Whether juries should be instructed on nullification is disputed: see *People* v. *Dillon*, 34 Cal. 3d 441, 490-3; 668 P. 2d 697, 728-30 (1983) (Kaus, J., concurring). Five states entrust even sentencing to the jury: Ark., Ky., Mont., Okla., Tex., Va.

[8] See Nelson, *supra*, at 165-71, tracing the transfer of law-finding functions from jury to judge in the first third of the 19th century. 'The basis for deciding cases suddenly shifted from personal and moral considerations to the formulations of broad economic policies and the determination of group interests. It is an account of the breakdown of the jury as an effective arm of administration because jurymen were unable to adjust their ways of thinking to legalistic conceptions of fair play, and of the consequent efforts of judges

manifestations of jury independence came to be condemned as improper. Criticism of the jury by more conservative spokesmen and especially by influential members of the legal profession multiplied and has not been stilled to this day.

Just as the criminal jury has been denounced in England and the United States by representatives of law and order, so the civil jury in tort litigation has become the darling of plaintiffs and the scourge of defendants. For the plaintiffs' bar the jury is the key to forensic success, jealously defended against any threat of retrenchment by means of organized propaganda and political action.[9] Even many defence attorneys prefer entrusting their case to a jury rather than to a judge, in the belief that it would be too risky to hinge the outcome on a single individual, often as likely to be suspected of pro-plaintiff bias,[10] and in the hope of persuading a sufficient number of the jury to stand their ground against the plaintiff.

In the following pages we will seek to find answers to this debate; on the one hand, in the opportunities which the procedural system and the substantive law offer juries to follow their own drummer and, on the other, in the controls available to keep them in bounds.

1.1   Size of jury and unanimity

The common law jury consisted of twelve veniremen whose verdict had to be unanimous. This tradition has been modified in both respects in many jurisdictions, in the Commonwealth[11]

---

to take matters into their own hands ... ': Malone, 'The Formative Era of Contributory Negligence', 41 Ill. L.Rev. 151, 152 (1946).

[9] *Infra*, ch. 5.

[10] The pervasive distrust of single judges, in contrast to the numerosity of juries, is a, perhaps the, major factor in the continuing professional support of the jury system. Multi-judge trial panels or a single professional judge with two lay associates, in use in some other countries, are foreign to common law procedure.

[11] e.g. an English Crown Court jury of 8 (majority verdict of 7); a New South Wales jury of 4 (majority verdict of 3). Respectively, Jury Act 1974 s. 17, Jury Act 1912 ss. 29, 66. In English courts other than Crown Courts, juries of at least 9 may render a verdict; the required majority being 10 out of 12 or 9 out of 10: ibid.

as well as in the United States.[12] The permitted size of the jury ranges down to six, and in some jurisdictions the jury may return a verdict by a majority of nine or ten.

The declared objectives of these reforms has been to promote administrative efficiency. Smaller panels and fewer 'hung juries' expedite adjudication. Additional claims that reduction in the size of juries does not materially affect their representative character or their verdicts have been disputed.[13] At all events, relaxation of unanimity would seem to favour plaintiff victories now that a defendant can no longer claim a mistrial by merely gaining one vote out of twelve. Even more favourable is the ruling in some states that there need not be an identity on all issues of the jurors constituting the necessary majority. It is thus possible for a plaintiff to get a favourable verdict with less than nine jurors in perfect agreement. For example, one or more jurors voting against liability may still make up the required majority on apportioning liability among plaintiff and defendant.[14] This abandonment of the necessity of consistency in special verdicts may expedite trials but at the cost of clarity and logic. Moreover, as implied in the US Supreme Court's decision that the verdict of less than six is incompatible with a

---

[12] The literature is collected in *Cooley* v. *Strickland Transportation Co.*, 459 F. 2d 779, 782 n. 9 (5th Cir. 1972). In federal courts, unanimity is still required (*Masino* v. *Outboard Marine Corp.*, 652 F. 2d 330 (3rd Cir. 1981)), but not a jury of 12 (*Colgrove* v. *Battin*, 413 US 149 (1973) upholding a local federal court rule of 6). The unanimity rule may induce defendants to seek a transfer to federal court.

[13] See Lempert, 'Uncovering "Nondiscernible" Differences: Empirical Research and the Jury-Size Cases', 73 *Mich. L. Rev.* 643 (1975); Kane, 'And Then There Were Twelve', 68 *Calif. L. Rev.* 1004 (1980).

[14] e.g. *Juarez* v. *Superior Court*, 31 Cal. 3d 759, 647 P. 2d 128 (1982). In *United Farmworkers* v. *Superior Court*, 111 Cal. App. 3d 1009, 169 Cal. Rptr. 94 (1980) of 11 jurors who attributed to the plaintiff 90% of responsibility, only 8 found that his negligence was a proximate cause at all! The 'same jurors' problem must be distinguished from 'patchwork verdicts'. The former concerns cases where the litigant must prevail on two or more essential issues, the latter where he may prevail for one or more different reasons, e.g. defendant's negligence may have consisted in speeding, dozing, or driving with bad brakes. Patchwork verdicts are generally tolerated. See Trubitt, 'Patchwork Verdicts, Different-Juror Verdicts, and American Jury Theory', 36 *Okla. L. Rev.* 473 (1983).

constitutional guarantee of jury trial,[15] there is clearly a limit
to the ongoing erosion of the traditional concept of jury trial.

1.2 Composition

Originally juries were selected for their knowledge of local
conditions; nowadays they are supposed to represent the
community at large. 'Represent' was for long understood to
imply 'speak for the community', not inconsistent with a
selection of 'right-thinking' citizens rather than a group
corresponding to, and including, various sections of the
pluralistic American polity. The practice of empanelling so
called 'blue ribbon' juries in certain cases was widespread and
constitutionally upheld.[16] Only towards the middle of our
century did protests about the class bias of the typical
middle-class jury begin to be heeded. Especially for disadvantaged criminal defendants fairness was suspect at the
hands of juries which contained no representation of ethnic
minorities and socio-economic groups outside the mainstream
of middle-America. Civil rights litigation, particularly over
desegregation, gave special impetus to the eventual constitutional requirement that jury selection not be systematically
biased in terms of racial or ethnic composition.

As a result, procedures for the selection of juries, civil as
well as criminal, have changed drastically. Jury pools must
now be representative in the sense of not excluding any
'recognizable group'.[17] Clear standards for selection and
qualification of jurors have taken the place of variant and

[15] *Burch* v. *Louisiana*, 441 US 130 (1979). That, like all other decisions
applying federal due process standards (14th Amendment) to state jury
requirements, dealt only with criminal trials.

[16] *Fay* v. *New York*, 332 US 261 (1947). See generally Van Dyke, *Jury
Selection Procedures: Our Uncertain Commitment to Representative Panels* (1977),
primarily focused on criminal trials. Cf. the 'special juries' in England,
abolished in 1949 (see Cornish, *supra*, at 33–5).

[17] e.g. Federal Jury Selection and Service Act of 1968 (28 USC § 1861–
9). A Uniform Jury Selection and Service Act (1970), based on the federal
pattern, has been adopted by some states. While this transformation was
prompted by criminal decisions, it applies equally to civil juries because their
rolls are usually the same.

ideologically tainted practices.[18] To that end, substantial efforts are now being made to rely not only on voter lists, which give inadequate representation to racial minorities and the young, but also on driving licences, tax returns, and welfare lists.[19] The only minimum qualifications are citizenship, local residence, and an ability to understand English. Nowadays few exemptions are allowed, in some states not even for lawyers and teachers.

1.3 Voir dire

Individual screening occurs at the *voir dire*.[20] A panel of, say, sixty veniremen, drawn from the general pool, is presented for ultimate selection of the twelve. The pretended purpose of the *voir dire* is to filter out bias and prejudice; in reality it is to seat a jury favourable to one's cause. Under the pretence of ensuring an impartial jury, it aims at a packed jury. If at one time justified to correct biased venire selection, peremptory challenges have become anachronistic since the adoption of modern safeguards to ensure a cross-section of the community. It is ironical that, after all initial efforts to ensure a jury representative of a heterogeneous society, we allow that representativeness to be later frittered away.[21]

While the practice is not uniform, it usually commences with questioning by the judge in order to eliminate those who are acquainted with any of the parties, likely witnesses, the lawyers, and persons associated with the aforementioned, or who are familiar with the circumstances of the case. In most state jurisdictions, however, the main drama of interrogation is then conducted by the parties' lawyers directly, in the belief

[18] e.g. the 'key man' or 'suggester' system, widely prevalent among federal (prior to 1968) and state courts, which sought out persons of 'good character, approved integrity, sound judgment and fair education'. See Friedenthal, Kane, and Miller, *Civil Procedure* § 11.10 (1985).

[19] See Kairys, Kadane, and Lehoczky, 'Jury Representativeness: A Mandate for Multiple Source Lists', 65 *Calif. L. Rev.* 776 (1977).

[20] *Voir dire* means speaking the truth. (*Voir* is a corruption of *verus*.)

[21] This observation by a Swedish observer is cited by Zeisel and Diamond, 'The Jury Selection in the Mitchell-Stans Conspiracy Trial', 1976 *A. B. F. Res. J.* 151, 174, a revealing analysis of the *voir dire* in a famous trial, illustrating many of the aspects discussed below.

that only unfettered, face-to-face, in depth interrogation can flush out prejudicial beliefs and bias.[22] It at once provides the lawyers with an opportunity to eliminate those perceived to be least sympathetic as well as to sell their case to the jury. Not only do the outlines of the case gradually emerge, but evidence that would not be admitted at the trial itself, such as that the defendant is insured, can be insinuated, as by legitimately questioning jurors on their relationship to persons employed in insurance. Recall the American lawyer who asked his English acquaintance when the trial commenced under English practice. The Englishman replied, 'When the jury is sworn.' The American was astounded: 'With us, that is when the trial is almost over.' Indeed, not only is the *voir dire* very time-consuming,[24] but the groundwork for victory may well be laid before the first witness is ever called. The importance attached to the process by trial lawyers is attested by the plethora of manuals, articles, and seminars devoted to the art.[25]

Challenges of individual jurors may be either for cause or peremptory. The former are for demonstrated bias or hostility,[26] the latter are entirely discretionary and therefore the

[22] Judge-conducted *voir dire* is more prevalent in federal courts, but also employed by some state judges. In California, the practice, though rarely used, survived challenge in *People* v. *Crowe*, 8 Cal. 3d 815, 506 P. 2d 193 (1973), but was quickly repudiated by amendment of Penal Code § 1078. For a spirited defence of lawyer-conducted *voir dire* see Babcock, 'Voir Dire: Preserving Its Wonderful Power', 27 *Stan. L. Rev.* 545 (1975).

[24] In some jurisdictions, often several days even in civil cases. The *voir dire* in the Los Angeles Hillside Strangler case took over 1 year; in the Mitchell–Stans trial 196 jurors were examined (Zeisel and Diamond, *supra*).

[25] Most of it is repetitive and of 'cookbook' quality. The Index of Legal Periodicals ('Jury') is a graveyard of this genre. More ambitious is Starr and McCormick, *Jury Selection: An Attorney's Guide to Jury Law and Methods* (1985). There is even an organization primarily devoted to jury selection, the National Jury Project, which maintains a staff of 15 and offices in Oakland, New York City, and Minneapolis, and publishes a loose-leaf manual. Numerous social science studies in the last 50 years, among them the high profile University of Chicago Jury Project, are summarized and critiqued by Simon, *The Jury: Its Role in American Society* (1980). These studies focus mostly on the criminal jury, use very small samples, and are quickly obsolescent and of greater interest to social psychologists than lawyers.

[26] e.g. California Code of Civil Procedure § 602, listing 8 grounds, including family relationship and conflict of interest.

more important. The number of challenges varies according to the jurisdiction. Peremptory challenges give a unique flavour to American trials at the onset.[27] In California, for example, each side has eight challenges in civil, twenty-six in criminal cases.[28] In federal courts, peremptory challenges in civil cases are limited to three.[29] This, together with the fact that the jury pool is drawn from a much wider venue, may significantly affect the composition of a federal jury. Litigants frequently have a choice between trial in state or federal courts. Their decision is usually influenced by tactical considerations among which the composition of juries is one. In tort cases, a more pro-plaintiff jury can usually be counted on in state court, especially in the larger cities. Other procedural factors no less influential are that, in contrast to federal practice, some states no longer require unanimous verdicts,[30] and that federal judges tend to 'run a tighter ship'.[31]

In criminal cases the effect of illicit bias is of course particularly prejudicial. Hence the prohibition of systematic exclusion of individuals on account of race or ethnic origin by peremptory challenges no less than in the selection of the jury pool.[32] In civil cases, however, lawyers on both sides also exert their energy and talents to seat a jury believed to be more sympathetic to their side than to the other. Some go so far as to employ jury consultants, specialists in social scientific methods, in survey research techniques as well as in research into the background of the jury pool for advance preparation.[33] But for the ordinary run most trial lawyers rely on stereotypes

[27] In England peremptory challenges in civil proceedings were denied as early as 1854 in *Creed* v. *Fisher*, 9 Exch. 472, and were limited to 3 even in criminal trials. (Criminal Law Act 1977 s. 43.) See the *Roskill Report* ch. 7 (1986) which recommended total abolition in criminal fraud cases. They are, however, not unknown in some Commonwealth jurisdictions: see e.g. Morison, *The System of Law and Courts Governing New South Wales* (2nd edn. 1984) § 19.08.

[28] Code of Civil Procedure § 601; Penal Code § 1070.

[29] 28 USC § 1870.

[30] *Supra*, n. 12.

[31] *Infra*, n. 104 (Jury Instructions).

[32] *Batson* v. *Kentucky*, 106 S. Ct. 1712 (1986); *People* v. *Wheeler*, 22 Cal. 3d 258, 583 P. 2d 748 (1978).

[33] See Simon, *supra*, ch. 3; 'Are Jury Consultants Worth the Cost?', *Nat. L. J.*, 21 July 1986.

that have become common coin in the profession, perhaps
fine-tuned by those who fancy themselves gifted with peculiar
psychological insights or intuitions.

1.4   Stereotypes

Pro-plaintiff bias in civil cases is for most purposes linked with
pro-defendant bias in criminal trials. Almost all imaginable
demographic characteristics are believed to affect jury bias:
ethnic origin, religion, marital status, sex, age, occupation,
wealth. For example, the emotional scale of ethnic groups,
and their corresponding receptivity to compassion, is ranked
from high to low: Irish, Jewish, Italian, Slavic; at the bottom:
English, Scandinavian, and German. The notorious Mr Belli
of San Francisco suffered much opprobrium on a recent
occasion for publicly stigmatizing Orientals as being un-
generous to plaintiffs, and stating that he would knock them
off juries first.[34] Then, again, Catholics and Jews are perceived
as compassionate and generous, unlike Protestants, especially
Lutherans; men as more sympathetic to victims, particularly
younger women; women are good for all defendants except
an attractive woman; persons in upper-income groups lack
concern for accident victims, while low-income individuals are
biased against corporate defendants. And so on.[35] As one
observer summarized:

The underlying values that seemed to be represented in the
characteristics cited above are ones usually associated with a minority
or majority ethnic or religious identification and an upper or lower
class identification. Jurors who identified with minority ethnic, racial
and/or religious groups are expected to believe that society's laws
are superimposed by members of the majority and that they may
be used to legitimate injustice. Such jurors are expected to empathize
with a party who is seeking redress of grievances for maltreatment

---

[34] In Belli's typically colourful language: 'The god-damned Chinese won't
give you a short noodle on a verdict. You've got to bounce them out of there.
In the last case that I tried, I used all my challenges getting rid of those sons
of the Celestial Empire.' But as the *National L. J.*, 16 August 1982 reported,
this advice to the Trial Lawyers at the ATLA convention apparently prompted the cancellation of his projected appearance at the ABA meeting in San
Francisco.

[35] See e.g. Penrod and Pennington, *Inside the Jury*, ch. 7 (1983); Simon,
supra.

especially if the complainant is a single individual and the negligent party a corporate body, and to have a generalized hostility towards what are defined as external symbols of authority. Finding in favour of a plaintiff and awarding higher amounts in damages is interpreted as an opportunity to express sympathy for an injured party and to demand a redress of grievances that carries with it the full backing of the legal structure.

Conversely, the following interpretations may be made about the meaning of a majority group identification. Such persons are believed to have more effectively internalized the values and norms that underlie the established legal code and to believe in their inherent justness. They are likely to have a stronger sense of identification with the established structure of authority, and a generalized respect for its external symbols. For such persons, the act of finding for the defendant or granting a lower award to the plaintiff, becomes an overt expression of their identification with the existing system of justice.[36]

Actually, such social science studies as have been made give only weak support to the perceived relationship between these stereotypes and their verdict preferences in typical cases. The only clear correlation is that based on juror identification with a particular party: jurors like their own kind, whether plaintiff or defendant. The famous University of Chicago jury project in the late 1950s set about for evidence to support the call, then fashionable among influential leaders of the legal profession,[37] to follow the English example of abolishing the jury at least in civil actions. Most of the empirical evidence, however, did not unequivocally bear out the critics' hunches; indeed it suggested that the correlation between jury and judge outcomes was remarkably close. On the other hand, the studies themselves have been fragmentary and are now rather dated. In any event, they do not contradict the still prevalent belief among trial lawyers that jurors tend to identify with one or the other party. In urban areas, in the 1980s, most jurors come from the lower income spectrum and would naturally identify with tort plaintiffs.[38] Of course, the fact

---

[36] Simon, *supra*, at 35–6.

[37] See, e.g., Frank, *Courts on Trial*, ch. 8 (1949).

[38] How can one explain that, according to 5 *Jury Verdict Research, Personal Injury Valuation Handbook* (1965), the recovery rate of professional people, as plaintiffs, is 25% below average, and they receive 17% smaller awards?

that verdicts go about equally to either party does not disprove this hypothesis: if the dice are loaded against defendants, or believed to be, that would influence their decision to settle rather than to litigate.[39]

1.5 Deep pockets

A common perception, prevalent especially among critics of the jury, is a bias against corporate and other defendants credited with deep pockets. In so far as that bias is reflected in excessive awards, this attitude may be an expression of a popular, latent hostility against powerful interests widely viewed as oppressors and exploiters of the common man.[40] If it also leads to greater readiness to find initial liability, it would be in train with the general tendency, shared also by a liberal judiciary, to convert the tort/insurance system to a general welfare function in the absence of any comprehensive social security programme. The deep pocket assures wide distribution of the cost and dispels any misgiving that the cost might be unaffordable. That it would ultimately be channelled into higher consumer prices and insurance premiums does not appear to be heeded, except in the case of motor-car liability, where a long-standing public relations campaign by the insurance companies has been an effective corrective to jury generosity.

These primarily anecdotal impressions have lately been supported by harder statistical information. A comprehensive

---

[39] See generally Priest and Klein, 'The Selection of Disputes for Litigation', 13 *J. Leg. Stud.* 1 (1984); Priest, 'Reexamining the Selection Hypothesis: Learning from Wittman's Mistakes', 14 *J. Leg. Stud.* 215 (1985). Plaintiff victories in San Francisco jury trials, 1960–80, hovered between 52% and 64%; in Cook County, Ill., around 52%. Almost consistently they were greater in the former city. But there were substantial disparities between different types of claim, varying from 33% and 41% for medical malpractice to 69% and 64% for worker injuries; these can be attributed to differential stakes more easily for the first than the second category (see Priest and Klein, at 37-41).

[40] For anecdotal evidence of such lower-middle class resentment see the jury interviews in *W. St. J.* 29 May 1986, at 1, 20.

study of civil jury cases in Chicago, 1959-79,[41] yielded the conclusion that, while deep pocket defendants were not more likely to be found liable unless plaintiffs had suffered very serious injuries, they usually paid larger awards than individual defendants when they were found liable. 'The awards that corporations paid to severely injured plaintiffs were 4.4 times those in similar cases against individual defendants, and 3 times those awards against government defendants. Where a government defendant might pay $75,000 and an individual defendant $50,000 to a severely injured plaintiff, a corporate defendant would pay $220,000.'[42] Even more pronounced is the trend of punitive damages: median awards in California against businesses were five times greater, on average nearly six times.[43] The difference between individual and corporate defendants reflects the fact that automobile cases account for the vast bulk of the first group.[44] This bears out the inference that the relative leniency shown to individual defendants may be due to greater empathy with fellow drivers as well as concern at the negative effect of high awards on insurance premiums. At any rate, in these cases the suspected jury tendency to use the tort system for redistribution of resources from rich defendants to unfortunate plaintiffs is not evident.

A connected trend is the notable shift in the mix of jury trials from motor traffic to products liability, malpractice, and government liability cases.[45] While automobile cases have largely become routine after many decades of experience, the

[41] Chin and Peterson, *Deep Pockets, Empty Pockets: Who Wins in Cook County Jury Trials* (1985). The study is based on the outcomes of 19,000 jury trials. Corporations made up slightly more than a quarter of all defendants. The bulk of these cases related to work injuries and product liability.

[42] Ibid. vii; see also 41-3 and Table 4.5. These figures were based on rigorous multivariate analysis.

[43] Peterson, Sarma, and Shanley, *Punitive Damages. Empirical Findings* 50-1 (1987). Also, the relationship between compensatory and punitive damages differs to the disadvantage of business: ibid. 61-2.

[44] Individual defendants made up 79% of automobile cases (ibid. 15). Reinforcing this impression is the fact that the trend since the 1970s of automobile awards has not shared the substantial rise in awards, at least for more severe injuries, in products liability and work injury cases (ibid.); Peterson, *Compensation of Injuries: Civil Jury Verdicts in Cook County* 55-6 (1984); Peterson and Priest, *supra*, at 54-7 (1982).

[45] The Chicago jury study by the Rand Corporation's Institute for Civil

larger stakes and lesser predictability of the others go to explain the greater reluctance to compromise and willingness on the part of plaintiffs to take a gamble on jury trial despite a less than even chance of victory.[46] It is perhaps ironic that juries are thus confronted more and more with cases for which they are less and less qualified. Increasingly issues are presented for legal adjudication which are far removed from the commonplace activities and accidents familiar to the man in the street. Not only does this new type of case necessarily call for evaluation of technical data and expert witnesses, straining sustained comprehension during lengthy trials; it also often presents disputes that are arguably unjusticiable because they are too many faceted for traditional legal adjudication. Such are the polycentric issues in typical product design-defect litigation to which reference was made in an earlier chapter.[47] If such issues overstrain the judicial function, as has been argued by some critics,[48] they are especially unsuitable for allocation to juries. This for the good reason that juries will tend to give insufficient weight to certain policy factors, such as cost and efficiency, as against safety and compensating injury, even assuming that on more general grounds one should countenance entrusting decisions of such wide-ranging and far-reaching social and economic welfare to a randomly selected group of unqualified citizens.

Justice, referred to in the previous 3 footnotes, revealed that, while automobile cases retained about 60% of the jury workload between 1960 an 1979, products liability and professional malpractice more than trebled mainly at the cost of worker injuries. Peterson and Priest, *supra*, at 12-16. The authors note that 'as the burden of the civil jury becomes more complex, concerns expressed by some about the competence of the lay jury become more pointed' (at 16).

In San Francisco the shift was even more pronounced. Automobile cases declined from 48% to 38%, while product liability trials increased from 4% to 10% and malpractice from 6% to 8%. Shanley and Peterson, *Comparative Justice, Civil Jury Verdicts in San Francisco and Cook Counties, 1959–1980*, s. 5.

[46] See *supra*, n. 39.   [47] Chapter 2.

[48] *Supra*, ch. 2, at n. 138. That some judges are equally susceptible is illustrated by a *New York Times* editorial, 27 Dec. 1986, excoriating a federal judge for assessing $5m. against a spermicide manufacturer in the teeth of an expert committee which, after reviewing 20 epidemiological studies, advised the FDA that the preponderance of evidence indicated no association between spermicides and birth defects.

Large awards against local government or other public entities have serious cost implications for the taxpayer. Fear of juries' insensitivity to this burden has persuaded some states, as a condition of waiving sovereign immunity, to impose dollar limits on awards,[49] others (including the federal government) to deny jury trial altogether.[50] Those that have not adopted protective measures, now suffer from the deep pocket syndrome, aggravated by a widescale withdrawal of insurance carriers from this line of business.[51]

2. JURY DISCRETION

Common wisdom has it that judges enunciate the law, juries apply it to the facts as they find them. This naïve definition of the jury function does not do justice to the reality of the legal process. For notwithstanding the traditional nomenclature, juries are not confined to questions of 'fact', in the sense of events that actually occurred or conditions that actually existed, but are called upon also to attach legal consequences to those facts.[52] The latter task involves normative

[49] Eight states have caps, ranging from $20,000 in Ky. to $100,000 in Ill. and NC. See *Civil Actions Against State Government: Its Divisions, Agencies, and Officers* (Shepard's/McGraw-Hill 1982). More are under way.

[50] Judicial Code (USC § 2402). 'The primary objection ... is that juries ... might tend to be overly generous because of the virtually unlimited ability of the Government to pay the verdict' (House Report No. 659, 1954 *US Code Cong. & Adm. News* 2716, 2718).

[51] This insurance 'crisis' has been specifically responsible for current attacks on the 'joint and several liability' rule. In California it is compounded by the constitutional inability of local governments to raise taxes (Proposition 13).

[52] These are sometimes referred to as 'questions of mixed law and fact'. At all events, jury functions cannot be ascertained by asking what is a question of fact; rather, if once determined to be a jury function, it is then designated as a question of fact. That the classification is influenced by legal policy is emphasized by the comparison between 'reasonableness' in negligence litigation, which is traditionally treated as a jury question, and 'reasonable and probable cause' as an element of malicious prosecution, which is classified as a question of law, and 'reasonable time' in commercial contexts, on which practice is divided. See Weiner, 'The Civil Jury Trial and the Law-Fact Distinction', 54 *Calif. L. Rev.* 1817 (1966); Harper, James, and Gray, *The Law of Torts*, ch. 15 (2nd edn. 1986). The classical article is by Thayer, '"Law and Fact" in Jury Trials', 4 *Harv. L. Rev.* 147 (1890).

judgments, giving concrete definition for the case in hand to such abstract standards as the reasonable man, foreseeability, and so on. Together with procedural rules which blur the effectiveness of the judge's instructions to the jury on the law to be applied, these aspects of the tort process in action open a wide door to 'jury law' through which prevalent street values can colour the outcome of litigation and in the longer run redefine legal rights and duties.[53]

The modern trend towards stricter standards of tort liability—towards 'negligence without fault'[54] or 'negligence in name only'[55]—has been observed in many countries.[56] That it transcends the division between civil and common law jurisdictions would suggest that it is a reflection of common, changing values in Western societies rather than specifically linked to peculiarities of the American legal culture. None the less, the extent of this transformation in most American jurisdictions far outpaces in distance and consistency that in other countries sharing similar social and economic values, including countries of the British Commonwealth. The jury's contribution to this dilution of the fault element and its effect on settlement practices has been a major element in this development.

2.1 Value judgments

The general verdict puts a powerful weapon in the hands of juries. Its 'featureless generality' permits them to announce a bare conclusion without indicating either grounds or reasoning: it is thus peculiarly result-directed. And as we shall see, this power is largely uncontrollable so long as the verdict falls

[53] It was one of the themes of Leon Green's work that the functions of judge and jury are closely interwoven and fluctuating in accordance with the need for judicial control depending on the type of case. See *Judge and Jury* (1930).

[54] Ehrenzweig, *Negligence Without Fault* (1951), repr. 54 *Cal. L. Rev.* 1422 (1966).

[55] Leflar, 'Negligence in Name Only', 27 *NYU L. Rev.* 564 (1952).

[56] e.g. for France see Viney, *Le Déclin de la responsabilité individuelle* (1965); Viney and Markesinis, *La Réparation du dommage corporel* (1985); for Germany see Weyers, *Unfallschäden* 81 ff. (1971); Kötz, *Deliktsrecht* 123-4 (3rd edn. 1983).

within an ever widening spectrum of judicial tolerance. There is no reason to suppose that the reasoning process of juries differs from that of the rest of the unschooled multitude, that is to say that it is primarily result-oriented, based on an amalgam of prejudice, impression, and intuition. Under the mask of a general verdict, juries are able (1) to fudge the facts so as to be able to apply the law to a desired result; (2) to make their own rules of law, either deliberately or through misunderstanding; or (3) to rush towards the desired outcome with 'brutal directness'—without either heeding the legal rules or determining the facts.[57]

The substantive law itself invites in many respects a wide-ranging normative discretion for juries. That the law of torts, especially the law of negligence, employs so many open-textured concepts or standards rather than precise and rigid rules[58] is by no means fortuitous. Most important, it thereby seeks to permit the infusion of community values and their adjustment over place and time. These objectives can of course be attained also by assigning those questions to professional judges, as has been done in England for most of this century. Leaving them to the jury, however, under the traditional common-law practice, is apt to increase greatly the opportunities for transforming the normative content of these concepts in accordance with the changing social values of the group from which juries are drawn.

In the United States, moreover, the jury also serves as the medium for accommodating the diversities of its people. More than in the homogeneous societies of Western Europe, in America tolerance of different values, reflecting ethnic, social, and religious diversity, has in modern times come to be hailed as a cardinal civic virtue. Thus, rather than striving for clear and consistent formulations of behavioural standards which can best be secured by authoritarian oracles of law, the

[57] See Frank, *supra*, at 111 (the 'realistic theory'). Contrast this setup with the Civil law procedure which sedulously observes a dissociation of the decision-maker from impressionistic judgment of the parties, witnesses, and other proof. There the 'judge of instruction' collects the evidence but the court (of which he may not even be a member) draws conclusions. See Merryman, *The Civil Law Tradition*, ch. 16 (2nd edn. 1985).

[58] Cf. Paton, *Jurisprudence* § 48 ('Principles, Standards, Concepts and Rules') (4th edn. 1972); Dworkin, *Taking Rights Seriously* 22–28 (1977).

ambiguities inherent in jury adjudication express the ambivalence of social and moral values demanded by a pluralistic society for its cohesion.[59]

This affirmative view of the jury's role has not gone without challenge. Justice Holmes, as early as 1881 in his celebrated book *The Common Law*,[60] asserted that to leave the question of negligence in every case, without rudder or compass, to the jury was a confession of the courts' inability to state a very large part of the law which they require the defendant to know and suggests by implication that nothing can be learned from experience. While laymen were initially more likely to possess the practical experience[61] for laying down intelligently the appropriate standard of conduct for a particular situation, it was ordinarily desirable (unless the standard was rapidly changing) to settle it once and for all instead of having juries oscillating to and fro. Thus in repetitive situations it would become unnecessary to take the jury's opinion again and again; indeed, 'the sphere in which [the judge] is able to rule without taking their opinion at all should be continually growing'.[62] Yet Holmes's attempt to entrench a 'stop, look, and listen' rule for level crossings[63] was eventually disowned by another master of the common law, Justice Benjamin Cardozo, who admonished on principle against framing standards of behaviour that amounted to inflexible rules of law.[64]

Not that Holmes's concern for adjudicative efficiency has remained isolated. It played a decisive role in the eventual disappearance of the civil jury from England and remains alive there in the tendency to endow decisions by trial judges on so-called questions of fact with quasi-precedential authority.[65] In America, it surfaces in the greater willingness

[59] Cf. Calabresi, *Ideals, Belief, Attitudes and the Law* (1985), which emphasizes these ambivalences, without however pinpointing the jury.

[60] At 123–4.

[61] Recall his aphorism that 'the life of the law has not been logic; it has been experience' (*The Common Law* 1 (1881)).

[62] Ibid., at 124.

[63] *Baltimore & Ohio R. R.* v. *Goodman*, 275 US 66 (1927).

[64] *Pokora* v. *Wobash Ry.*, 292 US 98 (1934). This contretemps had a parallel in England: see *Tidy* v. *Battman* [1934] 1 KB 319.

[65] Note, e.g., the standardization of deductions for non-use of seat belts in *Froom* v. *Butcher* [1976] QB 286.

by some judges to exercise their power of taking the case from
the jury.[66]

Perhaps more serious is an apprehension about insidious
jury encroachment on areas of legal policy. Not all judges
have yielded without protest to the latter-day judicial per-
missiveness in this respect. Justice Traynor, for example, one
of the most respected state court judges of our time, persistently
asserted the judicial prerogative of retaining control over
decisions of a policy nature. Such, for him, were questions like
the 'course of employment' under *respondeat superior*[67] and what
excuses to allow for violations of statute as a source of civil
liability.[68] His (unsuccessful) attempt to assert control over
damage awards reflected the same insistence on keeping the
jury within its traditionally allotted tasks.[69] More recently,
criticism has been mounting that the open-ended formulas of
negligence law permit the jury to inject not so much practical
experience as prejudice. Plaintiffs' lawyers know only too well
that an appeal to emotion and sentiment, to compassion
for victims, frequently overcomes the jury's mandate of
even-handedness between litigants. The increasingly wide
range of tolerance permitted to modern juries in passing on
questions like the defendant's blameworthiness, causation,
foreseeability of harmful consequences, or the plaintiff's contributory negligence has correspondingly reduced judicial
ability to enforce jury adherence to the law as laid down by
the judge. Instead, jury law seeps through the interstices of
black-letter law.

From the large pool of cases illustrating this tendency two
illustrations may suffice. In California foreseeability has for
most purposes become a sufficient test for 'duty' (as it has for
proximate cause), with the result that what was designed as
a control device to ensure judicial supremacy over sensitive
questions of policy has in effect been abandoned to juries.[70]

---

[66] *Infra*, n. 120.

[67] *Loper* v. *Morrison*, 23 Cal. 2d 600, 145 P. 2d 1 (1944).

[68] *Satterlee* v. *Orange Glenn School Dist.*, 29 Cal. 2d 581, 177 P. 2d 279 (1947)
(Traynor, J., concurring); vindicated in *Alarid* v. *Vanier*, 50 Cal. 2d 617, 327
P. 2d 897 (1958).

[69] *Seffert* v. *Los Angeles Transit Lines*, 56 Cal. 2d 598, 364 P. 2d 337 (1961).

[70] See e.g. 'This court has repeatedly eschewed overly rigidly common
law formulations of duty in favour of allowing compensation for foreseeable

Thus in *Bigbee* v. *Pacific Tel. & Tel.*[71] the plaintiff was injured
when a motor-car struck a telephone booth in which he was
standing and which was located fifteen feet from a main road.
The trial judge had granted summary judgment for the
defendants on the ground that they had no duty to protect
phone-booth users from the risk encountered by the plaintiff
and that neither their alleged negligence in locating the booth
nor any defect in the booth was a proximate cause of the
plaintiff's injuries, since he risk was unforeseeable as a matter
of law. He was reversed because 'in the light of circumstances
of modern life, it seems evident that a jury could reasonably
find that defendants should have foreseen the possibility of the
very accident which occurred here'.[72] As the dissent pointed
out, the decision 'will go far toward making all roadside
businesses insurers of safety from wayward travelers'.[73]

In certain types of cases pro-plaintiff judicial policy is
particularly conspicuous. For example, in one of the mass
immunization cases, the chance of infection from the Sabin
vaccine employed was $1:5.8m.$, that from the wild virus
rampant in the area $1:3,000$. The jury was permitted to find
that the child's infection stemmed from the former.[74] As one
study concluded, 'the courts have decided virtually every polio
case for the plaintiff on one or another theory'.[75] In other
areas a deliberate judicial policy to convert fault liability into
an instrument for distributive justice has led to the virtual
abandonment of all controls. Notable are claims by railroad
employees and maritime workers under federal legislation

injuries ... Rather than traditional notions of duty, this court has focused on
foreseeability as the key component necessary to establish liability': *J'Aire* v.
Gregory, 24 Cal. 3d 799, 805–6, 598 P. 2d 60 (1979). This was a case allowing
recovery for foreseeable economic loss sustained by a tenant as a result of the
defendant contractor's delay in completing repairs at the landlord's behest.
Cf. *Junior Books* v. *Veitchi Co.* [1983] 1 AC 520.

[71] 34 Cal. 3d 49, 665 P. 2d 947 (1983).

[72] Ibid., at 58.

[73] Ibid., at 61.

[74] *Reyes* v. *Wyeth Laboratories*, 498 F. 2d 1264 (5th Cir. 1974). The appeal
court would interfere neither with the trial judge's denial of a directed verdict
nor even with his refusal to grant a new trial.

[75] Franklin and Mais, 'Tort Law and Mass Immunization Programs:
Lessons from the Polio and Flu Episodes', 65 *Calif. L. Rev.* 754, 755 n. 3
(1977).

which has preserved employers' tort liability for work accidents in lieu of workers' compensation.[76] Not content with the total abolition of the defence of voluntary assumption of risk (1939) and the long-ago introduction of comparative negligence (1906, 1920), the US Supreme Court has insisted also on virtually unqualified access for plaintiffs to the jury, however weak their case.[77] This development was designed, and has resulted in an inevitable trend, to convert the employer's nominal tort liability into a system of no-fault compensation (without, however, its traditional quid pro quo of limited benefits).

In far fewer instances has legal doctrine been deliberately shaped by a desire to impose constraints on jury discretion. For example, liability for negligently causing mental disturbance (nervous shock) was long limited to actual impact and even today the conditions for allowing claims, if at all, by persons injured outside the zone of physical danger are more narrowly defined.[78] By comparison, the English progress occurred much earlier and without comparable concern over defining limits.[79]

So also reform of the law of occupiers' liability was long delayed in jurisdictions retaining jury trial, such as the United States and Australia, by concern over abandoning control over juries as the result of submitting the issue to the general standard of reasonable care. Whether or not originally designed for that purpose, the old regime by classifying entrants into the discrete categories of invitees, licensees, and trespassers, and attaching specifically defined standards to each, reduced the chance of juries' straying from their instructions and imposing a heavier burden on occupiers than the prevailing policy of the law contemplated. This approach of the law therefore served a somewhat similar function to the special verdict procedure (discussed below). In the United States the

[76] Federal Employers' Liability Act (FELA) 1906; Jones Act 1920. This development has been complemented by adoption of outright strict liability for unseaworthiness and the enormous expansion of that concept.

[77] *Infra*, n. 125.

[78] Prosser and Keeton, *The Law of Torts* § 54 (5th edn. 1984); Harper, James, and Gray, *supra*, at § 18.4.

[79] See Fleming, *supra*, at 145–52. Rather telling was the disclaimer in the latest of these extensions that damages would be in any event moderate: *McLoughlin* v. *O'Brian* [1983] 1 AC 410, 421.

reform of replacing it by a common duty of reasonable care, which in England had been effected by statute in 1957, has been mostly undertaken by the courts themselves. But to the extent that this has been adopted,[80] it was destined to have a much more radical effect than in England and other jurisdictions without jury trial.[81]

2.2   Contributory negligence

Contributory negligence has ever presented a challenge to juries' tolerance of a rule of law repugnant to the popular sense of fairness. That a negligent plaintiff should be barred from all recovery, particularly when he was much less to blame than the defendant, seemed a draconic and disproportionate sanction for carelessness. An exception was allowed when the defendant had the 'last clear chance' to avoid the accident, in which event the all-or-nothing rule dictated that the plaintiff could recover the whole amount of his loss. What the law did not countenance was an apportionment of damages corresponding to the parties' respective shares of fault, such as had become the rule of the civil law in most countries. Juries, however, frequently flouted their formal instructions on the law, applying in effect a double standard for plaintiffs and defendants especially in cases of severe injuries.[82] They could do so either by ignoring the plaintiff's fault entirely or by somewhat reducing the award and in effect apportioning damages. The latter method would become all too obvious

[80]   Prosser and Keeton, *supra*, § 62.

[81]   See, e.g., *Basso* v. *Miller*, 40 NY 2d 233, 352 NE 2d 868, 877 (1976) (Breitel J. concurring); Hawkins, 'Premises Liability After Repudiation of the Status Categories: Allocation of Judge and Jury Functions', 1981 *Utah L. Rev.* 15. The survival of jury trial also explains the reluctance of some Commonwealth jurisdictions to adopt a 'common duty' of care for occupiers: Fleming, *supra*, at 450.

[82]   See Harper, James, and Gray, *supra*, vol. iii, 392; Prosser, 'Comparative Negligence', 51 *Mich. L. Rev.* 465, 469 (1953), citing *Haeg* v. *Sprague, Warner & Co.* 202 Minn. 425, 430, 281 NW 261 (1938); Ulman, *A Judge Takes the Stand* 30-4 (1933). This tendency is sanctioned by Ariz. Constitution Art. 18 § 5: 'The defense of contributory negligence ... shall, in all cases whatsoever, be a question of fact and shall, at all times, be left to the jury.' It therefore precludes directed verdicts: see *Herron* v. *Southern Pacific Co.* (1931) 283 US 91.

when their award covered out-of-pocket loss alone, with no or inadequate allowance for perhaps severe pain and suffering. Some courts were prepared to tolerate such blatant exercise of jury 'nullification' of the law.[83] Preliminary studies since the official introduction of 'comparative negligence' (apportionment) in the early 1970s[84] appear to reinforce the impression that jurors were already applying it before the law was changed.[85]

The great majority of American comparative negligence statutes stipulate that, in order to recover any damages at all, the plaintiff's fault must not be more than the defendant's.[86] Should the jury be informed on the legal consequences of attributing a particular percentage of fault to the plaintiff? Courts which favour 'blindfolding' the jury reason that the jury's function is to find the facts, not to determine how the law should be applied to the facts.[87] In particular, jurors should not attempt to manipulate apportionment in order to achieve a result that may seem socially desirable to them.[88]

Whereas in the preceding situation disclosure of the legal consequences would favour plaintiffs, in the case of the 'joint and several liability' rule it might well have the opposite effect, once the jury understood that the least culpable of several tortfeasors might end up paying the whole bill; especially so if he happened to attract sympathy. Deep

---

[83] *Karcesky* v. *Laria*, 382 Pa. 227, 234; 114 A. 2d 150, 154 (1955) acknowledging the jury practice but affirming the trial judge's refusal to interfere with compromise verdict. For the doctrine of nullification, which has no legitimate role in civil cases, see *supra*, n. 7.

[84] See *supra*, ch. 2, at n. 56.

[85] Shanley and Peterson, *supra*, at 37–49.

[86] *Supra*, ch. 2, at n. 84.

[87] *McGowan* v. *Story*, 70 Wis. 2d 189, 234 NW 2d 325 (1975). Other courts have held that it was not reversible error to inform the jury: e.g. *Seppi* v. *Betty*, 99 Idaho 186, 579 P. 2d 603 (1978). See Comment, 'Informing the Jury of the Legal Effect of Special Verdict Answers in Comparative Negligence Actions', 1981 *Duke L. J.* 824.

[88] Statistical evidence suggests that this fear is unfounded. Awards are systematically smaller for negligent plaintiffs, regardless of the type of instruction, before being further reduced by the judge. Hammitt, Carroll, and Relles, 'Tort Standards and Jury Decisions', 14 *J. Leg. Stud.* 751 (1985).

pocket defendants, like public authorities, have therefore been advocating a mandatory disclosure of the rule to the jury.[89]

2.3 Damages

By far the widest scope for jury discretion is its power to assess damages. Its range of discretion here is less circumscribed than on issues of liability because concrete standards are lacking for ascribing monetary values to non-pecuniary losses and because judges have not been able or willing to communicate guide-lines that might have kept juries within a judicially approved range. As a result the American law of damages, particularly tort damages for personal injuries, is surprisingly unformulated, subject only to loosest post-verdict controls and in practice dispensed at greater variance with the few black-letter rules than any of the other jury functions. That the dollar amount awarded is of even greater importance to the plaintiff, and especially to his lawyer, than the decision on liability needs no complex demonstration: there is obviously a greater difference between a $30,000 and a $10,000 verdict than between a $10,000 one and a verdict for the defendant.[90] The ardent defence of the jury system by the plaintiffs' bar primarily reflects its financial stake in the hoped-for largesse from juries.

The most significant and interesting jury departure from 'law in books' is its tendency to relate damages to the defendant's degree of fault and financial resources. The authoritative rule is of course that ordinary damages are compensatory, not punitive, and therefore focused on the plaintiff's loss, not on the defendant's misconduct.[91] But

---

[89] *Kaeo* v. *Davis*, 719 P. 2d 387 (Hawaii 1986) held it proper to inform the jury.

[90] See Kalven, 'The Jury, the Law, and the Personal Injury Damage Award', 19 *Ohio St. L. J.* 158 (1958). This article reports tentative conclusions by its director from the University of Chicago Jury Project. The major work of that Project, Kalven and Zeisel, *The American Jury* (1966), is primarily focused on the criminal jury.

[91] Only in a minority of non-Common law countries does the degree of culpability (recklessness or intent) justify a different measure of damages: McGregor, *Int'l Encycl. Comp. L.* vol. xi ch. 9, 4–8.

American law has notably devoted little attention to precisely what nonpecuniary damages are supposed to accomplish, beyond the vacuous generality that the award should adequately compensate the injury. Is it to give 'satisfaction' as German law postulates[92] or is it to furnish 'substitute satisfactions' as protagonists of the 'functional approach' in British jurisdictions advocate?[93] Some critics have always suspected that a crucial element of non-pecuniary damages is to punish the defendant, if only to afford his victim the psychological satisfaction of seeing him punished.[94] American law at least condones the jury practice of hitching awards to the degree of culpability, quite apart from its open tolerance of distinct punitive damages for egregious wrongdoing.[95] In products cases where liability is strict, plaintiffs almost invariably still undertake the more arduous task of establishing fault, in the expectation that they will be rewarded by a correspondingly more generous award. They also resent the defendants' ploy of admitting liability and only contesting damages so as to exclude evidence of culpability that might have enhanced the award.[96]

The proclivity of juries to increase awards against deep pocket defendants has already been mentioned. Despite the clear legal policy against considering the financial resources of the litigants, reinforced by the prohibition against disclosure of insurance,[97] statistical evidence supports the impression that juries in practice tend to ignore these admonitions and on average award 4.4 times as much against corporate defendants

[92] See Stoll, *Int'l Encycl. Comp. L.* vol. xi, ch. 8 §§ 92–7.

[93] See Fleming, *supra*, at 215.

[94] e.g. Ehrenzweig, *Psychoanalytic Jurisprudence* § 207 (1971).

[95] See Kalven, *supra*, at 165. The 'proportionality principle', relating the extent of liability to the degree of fault, once enjoyed some support in European systems but has survived only as a residual Swiss provision, conferring a judicial discretion to reduce liability in cases of negligible fault: Stoll, *Int'l Encycl.Comp. Law*, vol. xi, ch. 8, 139–41.

[96] See *Fuentes* v. *Tucker*, 31 Cal. 2d 1, 187 P. 2d 752 (1947) (such evidence would be irrelevant to any disputed issue). See the outspoken protest by Carter, J., who thought it unfair to deprive the plaintiff by a ruse of a bonus that most plaintiffs enjoy in the ordinary run of cases.

[97] See Green, 'Blindfolding the Jury', 33 *Tex. L. Rev.* 137 (1954); Prosser and Keeton, *supra*, at 590–1.

as against individuals.[98] Other examples of the jury's 'polite war with the law' include reduction of damages on account of collateral benefits, a widow's prospect of remarriage and, as already mentioned, the plaintiff's contributory negligence under the old all-or-nothing rule. On the other hand, juries are widely suspected of increasing damages by making allowance for legal fees.[99]

Consistency and uniformity, as ideals of equal justice, are not valued to the same degree in American law as they are in England and many other countries. The very latitude inevitable in jury assessment of damages prompted English courts to abjure even occasional 'spot' jury trials, in the belief that the consistency attained by professional judges encourages settlements and avoids the American inflation of awards.[100] By the same token, the notorious generosity of American juries is the principal reason for the widely observed preference of many tort plaintiffs to seek an American forum rather than litigate at home.[101]

2.4 Jury instructions

Two features unique to the American process markedly contribute to the free-wheeling by juries. Both are germane to the manner in which the judge instructs the jury as to its tasks: the prohibition of judicial comment on the evidence and the formalistic manner of instructing the jury on the law to be applied by them to their findings of fact.[102]

[98] Chin and Peterson, *supra*, at 27.   [99] Kalven, *supra*, at 175-8.

[100] *Ward* v. *James* [1966] QB 273. Interestingly, the French position bears some comparison with the American, by treating the issue as one of fact and thereby giving considerable latitude to *juges du fond* (see Viney and Markesinis, *supra*, § 30).

[101] Forum shopping has been abetted by the liberal jurisdiction and choice of law rules of American courts. See *infra*, ch.7, n. 5. In the course of the Turkish Airline crash trial in Los Angeles a test case was tried by a jury. The English plaintiff had originally been offered $3,400 and had accepted $30,000 from the airline. He recovered a jury verdict of $1.5m. against the aircraft manufacturer! See Johnston, *The Last Five Minutes*, ch. 25 (1976).

English courts have condoned the choice of American forum to benefit from higher awards: *Castanho* v. *Root & Brown* [1981] AC 557.

[102] See generally Farley, 'Instructions to Juries: Their Role in the Judicial Process', 42 Yale L.J. 194 (1932); James & Hazard, *supra*, at § 7.14.

Under British procedure,[103] judges cultivate a collegial atmosphere in instructing the jury and, to this end, seek to assist them in as helpful a manner as their considerable communication skills allow in the evaluation of the evidence as well as in the application of law. Informality is lent to this phase of the trial by the extempore nature of the judicial advice. Not only do judges sum up the evidence but by analysing the testimony of witnesses and other proofs they also comment on their weight. More often than not they even express their own opinion, while making it clear that the jury will, of course, have the last word and ultimate responsibility for their findings. None the less, the judge's prestige assures that his charge will exert a strong, perhaps ineradicable, psychological effect on impressionable jurors, all the more so given the awesome atmosphere of the typical British courtroom and the social and educational gulf between judge and jury. As the judge's charge comes after the lawyers' summations, it is apt to overshadow their effect even more. Because of its lengthy agenda, the proceedings frequently last several hours.

American federal judges follow the British pattern in broad outline. But if they comment on the evidence, as many of them do, they do so generally with more restraint so as not to give the appearance of over-influencing the jury.[104] However, the greater dominance by the judge, added to the different composition of federal juries, marks a substantial difference between federal and state trials, reinforcing in turn the general preference by plaintiffs for the latter.

The situation in most state courts presents a strong contrast. Until the beginning of the nineteenth century, before a clearer line was drawn between fact- and law-finding functions, instructions to the jury were either not given at all or were brief and confusing.[105] Even later, as a legacy of the populist movement which was especially virulent in the expanding West, the balance between judge and jury remained heavily tilted toward the latter, so as to displace or at least modify

[103] Now mainly confined to criminal trials: see *supra*, n. 1.

[104] See James and Hazard, *supra*, at § 7.14.

[105] See Nelson, *supra*, at 26. The author notes that in Massachusetts a contributing factor was that cases were tried before a multi-judge panel where each judge, as well as each counsel, expressed his view of the law.

the élitist element represented by judicial authority. In consequence, statutes and even constitutions in many jurisdictions forbid judges to comment on the evidence, and some even require the judge's charge to precede the lawyers' summations. As a result, judicial control is greatly diminished; while at the same time less use is made of the traditional devices to prevent the case from going to the jury. The upshot is distinctly favourable to plaintiffs.

Jury instructions in American courts are highly formalized. Instead of being conveyed in colloquial language familiar to the ordinary juror, they are read out in stilted legalistic jargon tracking formulations of legal principle by appellate courts. There is reason to believe that these are not always clearly understood.[106] Thus even if the jury conscientiously wished to adhere to the black-letter law expounded by the judge, the verdict may well—unintentionally—express a deviant jury law. The artificiality and formalism of the instructions has been foisted on trial courts by the appellate practice of word-paring and readiness, when so inclined, to reverse for the slightest error. In view of the justified doubts concerning the jury's comprehension of the minutiae of the instructions, the appellate concern for strict verbal orthodoxy is almost bizarre.[107] Often, no doubt, prejudicial instruction is enlisted merely as a screen for reversing a verdict that seemed wrong on its merits, perhaps because of a suspected improper bias.

The litigants are entitled to request particular instructions; indeed their failure to do so or to object to such instructions as are actually given may disqualify them from challenging them on appeal. As a matter of practice, therefore, either

---

[106] For a detailed analysis of one asbestos case, which bears out this thesis, see Selvin and Picus, *The Debate over Jury Performance* (1987). 'The jurors usually are as unlikely to get the meaning of those words as if they were spoken in Chinese, Sanskrit or Choctaw': Frank, *supra*, at 116. Farley, *supra*, at 214, adds: 'But it is impressive to the public and it clothes the jurors with a sanctimonius mantle of enlightenment which gives them a sense of peace and accord with authority.' For Green (*Judge and Jury* 351 (1930)), a sympathetic observer, 'for the most part they are ritual'. California juries are now entitled to request written copies of instructions: Code Civ. Proc. § 612.5.

[107] The 'harmless error' rule (*infra*, n. 114) would provide a well-justified reason for ignoring most glitches.

party will usually submit written instructions, carefully drafted with a favourable slant, on which the judge may draw for his final formulation. In most jurisdictions, however, instructions are nowadays derived mainly from 'patterns' drafted by semi-official sources, such as California's BAJI (Book of Approved Jury Instructions).[108] These model instructions are mostly based on verbatim formulations of legal principle by appellate courts, frequently modified in the light of the latest rulings, sometimes rulings specifically on a particular model instruction. They were strongly encouraged and came into use in the post-World War II period as a signal improvement on the poorly drafted and poorly delivered instructions formulated by individual judges.[109] Even so, while they may have raised the general quality and reduced the number of appeals,[110] they compare very unfavourably with the British model as a means of communication. The most serious problem is that they tend to be unduly stylized and abstract, untailored to the concrete circumstances of the case in hand, and therefore make undue demands on comprehension by an untutored jury.[111] Encouraged by several social science investigations, an effort to develop simplified jury instructions is at last

[108] California was the pioneer of pattern instructions with the first edition of BAJI in 1938. The project was undertaken and continues to be sponsored by the Superior Court of Los Angeles County.

[109] e.g. Joiner, *Civil Justice and the Jury* 83 (1962) ('On the whole, judges have done a very poor job of instructing').

[110] *BAJI California Jury Instructions: Civil*, viii (6th rev. edn. 1977), quoting Chief Justice Traynor's Foreword to the 5th edn. Chief Justice Gibson (1940–64) is especially credited with pushing the use of BAJI in California.

[111] Take for example the following BAJI instruction:

BAJI 4.00
RES IPSA LOQUITUR—NECESSARY CONDITIONS FOR APPLICATION

On the issue of negligence, one of the questions for you to decide in this case is whether the [accident] [injury] involved occurred under the following conditions:

First, that it is the kind of [accident] [injury] which ordinarily does not occur in the absence of someone's negligence;

Second, that it was caused by an agency or instrumentality in the exclusive control of the defendant [originally, and which was not mishandled or otherwise changed after defendant relinquished control]; and

Third, that the [accident] [injury] was not due to any voluntary action or

getting under way.[112] Moreover, 'continuing education of the bench' programmes give particular attention to training judges to read instructions in a way that communicates rather than bores.

Another difficulty is that the instructions come at the end of the trial so that the jurors are in the dark throughout the taking of evidence as to what they will eventually have to decide. Many judges now give 'mini' instructions early on and during the course of the trial to assist jury orientation.

contribution on the part of the plaintiff which was the responsible cause of his injury.

BAJI 4.02 (1970 Revision)
RES IPSA LOQUITUR—WHERE ONLY A PERMISSIBLE INFERENCE OF NEGLIGENCE

From the happening of the [accident] [injury] involved in this case, you may draw an inference that a [proximate] [legal] cause of the occurrence was some negligent conduct on the part of the defendant.

However, you shall not find that a [proximate] [legal] cause of the occurrence was some negligent conduct on the part of the defendant unless you believe, after weighing all the evidence in the case and drawing such inferences therefrom as you believe are warranted, that it is more probable than not that the occurrence was caused by some negligent conduct on the part of the defendant.

In order to put these instructions into context, Evidence Code § 646 may be helpful:

(c) If the evidence, or facts otherwise established, would support a res ipsa loquitur presumption and the defendant has introduced evidence which would support a finding that he was not negligent or that any negligence on his part was not a proximate cause of the occurrence, the court may, and upon request shall, instruct the jury to the effect that:

(1) If the facts which would give rise to a res ipsa loquitur presumption are found or otherwise established, the jury may draw the inference from such facts that a proximate cause of the occurrence was some negligent conduct on the part of the defendant; and

(2) The jury shall not find that a proximate cause of the occurrence was some negligent conduct on the part of the defendant unless the jury believes, after weighing all the evidence in the case and drawing such inference therefrom as the jury believes are warranted, that it is more probable than not that the occurrence was caused by some negligent conduct on the part of the defendant.

[112] See Perlman, 'Pattern Jury Instructions: The Applications of Social Science Research', 65 *Nebr. L. Rev.* 520 (1986), reporting on the newly drafted Alaska civil jury and federal criminal jury instructions (with samples).

To summarize, jury instructions serve multiple purposes. Besides their primary function of enlightening the jury on the law, they offer a method by which appellate courts control juries and trial judges.[113] Not that appellate courts consistently exercise this important power, which could bring the judicial business to a standstill. They in fact habitually close an eye to the ever-present errors (except in serious criminal cases) by either ignoring complaints, denying leave to appeal, or excusing 'harmless errors' not entailing a miscarriage of justice.[114] By the same token, the power can also be wielded with a view to promoting distinct policies, such as to nullify capital punishment or to assist plaintiffs' recovery.[115] The discretionary use is typical of the subtle balances of the common law method in tort litigation.

3. JURY CONTROLS

Over time various procedural controls developed to set boundaries beyond which juries were not permitted to stray. In their heyday, this resulted in 'a uniquely subtle distribution of official power, an unusual arrangement of checks and balances'.[116] But, like other aspects of the judicial process,

[113] Farley, *supra*, at 195, adds two other functions: 'From the Lawyer's View Point as Traps for the Courts', and as 'a Method by which the Trial Court Maintains its Integrity'. The last addresses so-called cautionary, as distinct from peremptory, instructions.

[114] Admonition against reversal for harmless errors is often enshrined in statute or even constitution. Thus California Constitution Art. VI § 13: 'No judgment shall be set aside, or new trial granted, in any cause, on the ground of misdirection of the jury, or of the improper admission or rejection of evidence, or for any error as to any matter of pleading, or for any error as to any matter of procedure, unless, after an examination of the entire cause, including the evidence, the court shall be of the opinion that the error complained of has resulted in a miscarriage of justice.' In California, the practice of reversing judgments for prejudicial error, especially when prejudicial to plaintiffs, continues unabated despite the mandate of its Constitution Art. VI § 13. See MacLeod, 'The California Constitution and the California Supreme Court in Conflict Over the Harmless Error Rule', 32 *Hast. L. J.* 687 (1981). Disregard of the rule in relation to criminal convictions has led to protests by the conservative minority on the court.

[115] Both policies have been charged against the California Supreme Court.

[116] Kalven and Zeisel, *supra*, at 498. Among these are also rules of evidence,

relaxation of these controls has had a profound effect in shifting the practical operation of the tort system from a pro-defendant to a pro-plaintiff orientation. Since these controls are triggered by largely subjective judgments of trial and appellate judges, this transformation has transpired less in changes of articulated doctrine than in the gradual modification of judicial practice.[117] As already explained at length, the progressive 'liberalization' of tort law is in larger measure the result of abandoning judicial control over juries than the consequence of reforming legal 'rules'. Thus the formal continuity of law remains preserved while its substance is being transformed.

American civil procedure offers both pre- and post-verdict devices for judicial control of juries, which significantly transcend those inherited from English practice.[118] The former are specifically addressed to questions of whether the plaintiff can 'get to the jury' at all. He must pass two major thresholds: nonsuit and directed verdict.[119] Both turn on the sufficiency

such as the hearsay rule, based on doubts of juries' capacity to weigh evidence judiciously. The lesser importance, or non-existence, of such restrictive rules in other legal systems (also their decline in England) is usually attributed to their having professional judges. Most of these rules are not peculiar to tort trials; one exception is the rule against proof of subsequent repairs to prove the dangerous condition or defective design (e.g. Calif. Evidence Code § 1151).

[117] Even if we had statistics of directed verdicts over a substantial period, they would not indicate the extent to which any change of practice affected plaintiffs' decision to risk a jury trial. In the Chicago Jury Study, between 8% and 15% of cases terminated in a directed verdict for the defendant: Chin and Peterson, *supra*, at 24.

[118] See James and Hazard, *supra*, at § 7.12-22. The constitutionality of the directed verdict has been upheld on the ground that there were common law analogues by which a judge could determine that the evidence was insufficient (demurrer to the evidence, orders of a new trial): *Galloway* v. *US*, 319 US 372, 390 (1943). But except for demurrer to the evidence (which was very risky to plaintiffs), these resulted only in orders for a new trial, not judgment.

[119] Demurrers to the pleading precede the empanelling of the jury, for if successful, they connote that even if all the facts alleged are proved, they do not disclose a cause of action. (Demurrers to the evidence, *supra*, n. 118, are now obsolete.) So does summary judgment, a modern device, when on the basis of affidavits there is no genuine issue of material fact and the moving party is entitled to judgment as a matter of law. See generally, Friedenthal, Kane, and Miller, *supra*, ch. 9 ('Adjudication Without Trial').

of evidence. At the end of the plaintiff's case, the defence will usually move that the plaintiff be nonsuited because his proof would not justify a jury reasonably to find for him. If, in the judge's opinion, a jury could reasonably find either way, the motion must be denied.[120] At the end of his own case, the defendant may ask for a directed verdict on the ground that, having regard to all the evidence, a jury would not be justified to find for the plaintiff.

If this is also denied and the jury actually finds for the plaintiff, the defendant may yet spike the plaintiff's victory by moving for judgment notwithstanding the verdict.[121] The conditions for granting this motion are identical to those for a directed verdict, but postponement may be preferable so as to avoid the necessity of a new trial in case an appeal court disagrees with the judge's ruling. Besides, if the jury actually returns a verdict as the judge thinks it ought to, the judgment based on that verdict will be practically invulnerable to appeal.

Rather than setting aside the jury's verdict by entering judgment for the defendant, the trial judge may order a new trial. Because the effect is much less drastic, the judge is generally given greater discretion. In some states, his discretion is unfettered, the judge in effect acting as a 'thirteenth juror' if dissatisfied with the verdict for one reason or another.[122] In others, the trial judge or appellate court may order a new trial if the verdict is against the (clear) weight of the evidence. In yet others, the 'federal test' requires the judge to accept the jury's findings on matters of credibility of testimony and weight of evidence, but allows him to disagree if he 'clearly see[s] that they have acted under some mistake or for some improper motive, bias or feeling'.[123]

[120] If the motion is granted, it may be with or without prejudice. If the former it is the equivalent of a directed verdict and often so-called. Until the modern period, nonsuit was always non-prejudicial, allowing the plaintiff another try. Today, the choice hinges on whether there is any possibility of a successful second try. A directed verdict is of course always prejudicial.

[121] Or judgment n. o. v. (*non obstante veredicto*).

[122] The California trial bar has endeavoured to tame this power by procuring a requirement of elaborate written findings in support of such order: Code of Civil Procedure § 657.

[123] *Aetna Casualty & Surety* v. *Yeatts*, 122 F. 2d 350, 353 (4th Cir. 1941).

The standards for invoking these judicial controls, especially those involving a denial of jury verdict, have occasioned much controversy. Those with an ideological or professional stake in jury trial, principally plaintiffs' lawyers and civil rights advocates, condemn them as 'jury subornation through judicial decisions';[124] defence lawyers look upon them as necessary protection against jury tyranny. Both the formulation of standards, and even more, their application have been inevitably influenced by policy preferences. The clearest illustration, previously mentioned, is found in the context of FELA litigation concerning claims by injured railway employees.[125] Under the close direction of the US Supreme Court, the applicable negligence standard has been transformed virtually into a workers' compensation system, but with common law damages, by the dual strategy of permitting juries to find for plaintiffs on the basis of 'any' evidence of 'any' negligence.[126] For example, when the evidence was equally poised between two hypotheses to account for a switchman's death—one, that he had been struck by a mail hook from a passing train; the other, that he had been killed by a hobo, many of whom frequented the area—the plaintiff's verdict was upheld.[127] In the Court's view

It is no answer to say that the jury's verdict involved speculation and conjecture. Whenever facts are in dispute or the evidence is such that fair-minded men may draw different inferences, a measure of speculation is required on the part of those whose duty it is to settle the dispute by choosing what seems to them to be the most

[124] *Galloway* v. *US*, 319 US 372, 404 (Black, J., dissenting).

[125] The Act (46 USCA §§ 51–60) was originally passed in 1906 and has the continued support of organized labour. The dilution of the negligence standard and the prospect of common law damages has made it more attractive than standard workers' compensation. From its inception comparative negligence (apportionment) applied in lieu of the common law defence of contributory negligence, and the defence of voluntary assumption of risk was abolished in 1939. See Symposium in 18 *Law & Contemp. Prob.* (1953).

[126] The 'any negligence' rule was established in a series of post-war cases from *Tiller* v. *Atlantic Coastline R. Co.*, 318 US 54 (1943) to *Rogers* v. *Missouri Pacific R. Co.*, 352 US 500 (1957) and *Gallick* v. *Baltimore & Ohio R. R. Co.*, 372 US 108 (1963).

[127] *Lavender* v. *Kurn*, 327 US 645 (1946); so also *Rogers* v. *Missouri Pacific R. Co.*, *supra*.

reasonable inference. Only when there is a complete absence of probative facts to support the conclusion reached does a reversible error appear.[128]

Whether this relaxed FELA standard should apply to all federal trials remains controversial.[129] In state courts a variety of verbal standards prevails,[130] but no one would question that the general trend over this century has been decidedly to favour tort plaintiffs through the increasing reluctance on the part of trial judges, and even more noticeably of appellate courts, to interfere with the jury role.[131]

This tendency is reflected not only in changing formulations of the standard for judicial intervention (like the just-mentioned FELA standard) but also in developments of rules of evidence and proof. Most conspicuous in the latter regard is the liberalization of *res ipsa loquitur* by means of which a jury may find negligence notwithstanding the absence of direct proof of causation or fault. Both the conditions for triggering this device, particularly the requirement of exclusive control, and its procedural consequences have become steadily more favourable to plaintiffs. Thus what began as an aid to establishing merely a prima-facie case, at most permitting a jury to infer nefligence,[132] has many jurisdictions developed into a presumption affecting the burden of producing evidence, if not actually reversing the burden of proof.[133] Miraculously, a weak case became transformed into a strong case, from merely enabling a plaintiff to 'get to the jury' to even giving him the benefit of a directed verdict in appropriate cases. Thus, under the American system, rules of evidence immediately

[128] Ibid., at 653.

[129] See Johnston, 'Jury Subordination Through Judicial Control', 24:4 *Law & Contemp. Prob 24 (1980)*.

[130] The various standards for directed verdicts are listed in *Pedrick* v. *Peoria & Eastern R. R. Co.*, 37 Ill. 494, 229 NE 2d 504 (1967).

[131] Malone, 'The Formative Era of Contributory Negligence', *Essays on Torts*, ch. 4 (1986) documents judicial efforts to counteract jury bias against railroads.

[132] *Byrne* v. *Boadle* (1963) 2 H. & C. 722.

[133] See Prosser and Keeton, *supra*, at § 39; Harper, James, and Gray, *supra*, at §§ 19.5–12; *supra*, n. 111. England has succumbed to the same temptation, though its significance is much less under a system without jury: *Moore* v. *Fox*, [1956] 1 QB 596 (CA); see Fleming, *supra*, at 300.

translate into defining and redefining the division of functions between judge and jury.

3.1 Damage awards

As previously explained, in tort actions the jury's discretion in fixing damages is even wider than that in determining liability. While the judge's instructions are mired in vacuous generalities, the jury is presented with extravagant demands by the plaintiff's attorney.[134] The upward pressure which has more than doubled the real value of jury awards in tort actions since 1960[135] reflects the growing success of the plaintiffs' bar in pursuing not only their client's interest but also their own in so far as their fees are linked to the award.

The proclivity of juries to empty deep pockets has been repeatedly noted. But judicial efforts to blindfold the jury from illicit knowledge have been largely thwarted. When not obvious from the name of the defendant, it can be surreptitiously conveyed in the course of the *voir dire* or even explicitly brought out under the ruse of claiming punitive damages.[136] The continuing prohibition of 'direct actions' against liability insurers, even where insurance is compulsory, in order to insulate the jury from the forbidden knowledge, has become an idle pretence, a reminder at best of more innocent days. Post-verdict control is therefore all the more necessary but also clumsy and difficult. Since, unlike liability, the question is not reducible to either/or, the judge cannot directly substitute his own for the jury's determination of the proper amount. In the last resort, the only corrective against an excessive or inadequate award is therefore a new trial. In deciding whether the award falls within the acceptable range, the judge is himself thrown back on a discretionary judgment,

[134] Many jurisdictions tolerate the 'per diem' argument, allotting a specific value for each day of suffering. See *Seffert* v. *Los Angeles Transit Lines, supra,* at 513, 364 P. 2d at 346. Nor have proposals to eliminate the '*ad damnum*' clause of the pleadings (DRI, *Responsible Reform* 25 (1969)) met with any success.

[135] See Shanley and Peterson, *supra,* at 26–30.

[136] The proposal of a bifurcated procedure under which the jury would first determine liability and whether to award punitive damages before admitting evidence of the defendant's wealth has been incorporated in Sen. Danforth's products liability reform bill (s. 1999).

aided however by a knowledge of general trends undistorted by lay perceptions gained from media reports of spectacular awards. That range, however, is very wide indeed, and variation between judges adds to the unpredictability of judicial response.

The problem is aggravated in dealing with awards of punitive damages where comparison with amounts awarded in other cases is not considered relevant. Instead, the award must be examined in the light of the circumstances of each individual case, having regard to such particular factors as the defendant's wealth, the degree of blameworthiness of his conduct, and the potential deterrent effect of the award.[137] All the same, a high percentage, particularly of the highest awards, are being substantially reduced.[138] In the leading California products liability case,[139] the jury's award of $125m. punitive damages against a wealthy automobile manufacturer was remitted to $3.5m. This discrepancy by a factor of more than thirty illustrates the subjectivity of assessing these damages; nor did the affirmance of the judge's $3.5m. suggest that the appeal court would not also have been prepared to condone $10m., $1m., or $350,000.

All jurisdictions agree in allowing a wider discretion to trial judges than to appellate courts in their review of damage awards. A common formula for the latter is that the verdict must not be so large 'that, at first blush, it shocks the conscience and suggests passion, prejudice or corruption on the part of the jury'.[140] Far more discretion is allowed to the trial judge, who is in a better position to form an opinion on the fairness of the verdict in the light of the general conduct and atmosphere of the trial.

---

[137] *Infra*, ch. 6, at n. 134.

[138] In California almost $\frac{1}{2}$, the final payment, on average, being about 50%. But most reductions resulted from settlements rather than *remittiturs*. Peterson, Sarma, and Shanley, *supra*, at 28–30 (1987).

[139] *Grimshaw* v. *Ford Motor Co.* 119 Cal. App. 3d 757, 174 Cal. Rptr. 348 (1981). In view of the extravagant jury award which the trial judge evidently deemed so grossly disproportionate as to suggest that it was the product of passion or prejudice, did it not also raise a suspicion that it tainted the verdict on liability so as to justify a retrial on all issues?

[140] *Seffert* v. *Los Angeles Transit Lines, supra*, at 507, 364 P. 2d at 342.

For the sake of efficiency, and notwithstanding constitutional challenges for invading the province of the jury,[141] a widespread practice has developed of offering to the litigants a judicial award as an alternative to a new jury trial. A *remittitur* offers the plaintiff a smaller sum than the excessive jury verdict, an *additur* offers the defendant a larger sum in lieu of an inadequate award. There is a division as to how the amount of either order should be determined. According to one view, which is least intrusive on the jury's constitutional function, a *remittitur* should be fixed at the maximum amount that a court would allow to stand (or the minimum in the case of an *additur*); according to another, the trial judge is authorized to exercise his own independent judgment of what would be fair and reasonable.[142]

Notwithstanding the common-law tradition of regarding a jury verdict in a damages action as indivisible, orders for a new trial limited to damages have long been recognized as permissible 'if the issue to be retried is so distinct and separate from the others that a trial of it alone may be had without injustice'.[143] The problem is that an inadequate award usually suggests a compromise verdict, in which case it would be unfair to the defendant to reopen the issue of damages against him while letting the verdict on liability stand. Some courts regard a seriously inadequate verdict as practically conclusive evidence of an improper compromise, for example a verdict limited to out-of-pocket expenses in a serious personal injury case. Others, however, are prepared to condone compromise verdicts as an expression of community values, if there is no suspicion of improper prejudice or mistake. Such used to be particularly prevalent in cases involving contributory negligence when juries in effect applied a rule of comparative

---

[141] *Additurs* were condemned by the US Supreme Court in *Dimick* v. *Schiedt*, 293 US 474 (1935), as in violation of the 7th Amendment. But *remittiturs* were upheld in *Dorsey* v. *Barba*, 38 Cal. 2d 350, 240 P. 2d 604 (1952) as less intrusive on the jury's constitutional role. All the same, a good many courts disavow *remittiturs*, content with the trial judge's power to order a new trial (which would be exercised more sparingly). See *Firestone* v. *Crown Center Redevelopment Corp.*, 693 SW 2d 99 (Mo. 1985) (en banc).

[142] James and Hazard, *supra*, at 399.

[143] *Gasoline Products Co.* v. *Champlin Refining Co.*, 283 US 494, 500 (1931).

fault rather than the official all-or-nothing rule.[144] What is much more difficult to defend is a judge's denial in such circumstances that the verdict represented an improper compromise and his order of a new trial limited to damages: 'it would let the plaintiff have the best of both possible worlds.'[145]

3.2 Special verdicts

Special verdicts or interrogatories can serve a useful function in keeping juries loyal to their instructions and are therefore welcomed alike by critics and supporters of the jury system.[146] Ordinarily, as noted, juries return a general verdict. But judges may choose to submit specific questions on different issues, such as: Was the defendant negligent and in what respect? Was his negligence, if any, a proximate cause of the plaintiff's injury? Was the plaintiff guilty of negligence? Did it contribute to the injury? And so forth.

The procedure has several possible advantages. One is to make the jury's task more comprehensible particularly in complex litigation. A second is to prevent juries from manipulating the outcome of the action, as in cases of comparative negligence by requiring specific findings of the total loss and separate allocations of fault to the parties in terms of percentages, leaving it to the judge to translate them into dollars and cents.[147] A third is to discourage compromise and patchwork verdicts.[148] Finally, it minimizes the risk that an erroneous instruction or other procedural flaw has necessarily tainted the verdict, because it may be shown to be irrelevant in view of a specific finding by the jury that would in any event have supported the verdict rendered. Thus a false instruction on the standard of care would not be prejudicial

---

144 *Supra*, n. 82.

145 James and Hazard, *supra*, at 402.

146 Ibid. § 7.15.

147 'The [whole] thought behind the special verdict [was] to free the jury from [anything] which would inject the feeling of partisanship in their minds.' *Seppi* v. *Betty*, 99 Idaho 186, 191; 579 P. 2d 683, 688. See also *supra*, n. 87 on 'blindfolding' the jury as to the consequences of apportionment, pursuant to the same policy.

148 See Trubitt, *supra*, at 496–502.

if the jury specifically found that the negligence was not a substantial cause of the injury.

In origin a device for jurors, *sua sponte*, to avoid the risk of attaint, special verdicts have thus been transformed into a means for wresting control over the outcome of litigation from the jury. Before the 1930s they were regularly used only in Wisconsin, Texas, and North Carolina, but were then adopted by the Federal Rules.[149] They became routine with the adoption of comparative negligence.

3.3 Bifurcated trials

It is understandable for many jurors to be influenced in favour of the plaintiff by the gravity of his injuries. In order to mitigate such a spill-over of sentiment from damages to liability, the trial could be split into two separate phases to keep the jury in the dark about the injuries while pondering the issue of guilt. Bifurcated trials are now widely authorized, but their primary purpose is to save court congestion rather than to obviate prejudice. Their use for the latter purpose, while permissible if not employed routinely,[150] is of course as much deplored by plaintiffs as it is welcomed by defendants.[151]

---

[149] See Dudnik, 'Special Verdicts: Rule 49 of the Federal Rules of Civil Procedure', 74 *Yale L. J.* 483 (1965).

[150] *Morley* v. *Superior Court of Arizona*, 131 Ariz. 85, 638 P. 2d 1331, 1333 and cases collected in 78 ALR Fed. 891, §§ 6, 11. Exclusion of the plaintiff from the liability phase is also permitted in some jurisdictions if his appearance would be prejudicial and he cannot understand the proceedings and assist his counsel: *Helminski* v. *Ayerst Lab.* (6th Cir. 1985) 766 F. 2d 208.

[151] Criticized by Weinstein, 'Routine Bifurcation of Jury Negligence Trials', 14 *Vand. L. Rev.* 831 (1961); advocated by Mayers, 38 *U. Pa. L. Rev.* 389, 393–5 (1938).

5

THE GLADIATORS

The chief 'players' in the handling and disposition of tort claims are lawyers and insurers. This chapter will speak about their organization, their motivations, and their influence on the decision-making process in court and legislature.

Specialization by subject area is a common trait among lawyers practising in a complex legal system like the American.[1] The law of torts has not remained immune to this tendency. While the odd personal injury claim may remain a staple of general law practices, especially in small communities,[2] most of the more successful tort lawyers practice exclusively in this field. One of the reasons is that the practice is pre-eminently a trial practice before juries, which calls for peculiar talents and skills. Another is the expertise required these days to cope with technical and complex evidence in products liability and medical malpractice actions, which have become the main fare of profitable litigation. Indeed, specialization has progressed further into specific types of claims, like aviation or medical negligence.

But the most peculiar feature of the American legal profession is the chasm dividing the plaintiffs' and defendants' bar in tort, especially personal injury, litigation. This division reaches back to the middle of the nineteenth century when a new type of lawyer, burdened with social and ethnic handicaps, began filtering into the profession and gravitated to causes on behalf

[1] See Johnstone and Hopson, *Lawyers and Their Work* 134–57 (1967). Specialization is now encouraged: see ABA Standing Committee on Specialization (*Handbook on Specialization*, 1983); Greenwood and Frederickson, *Specialization in the Medical and Legal Professions* (1964).

[2] A study covering Massachusetts lawyers in 1970, prior to automobile no-fault, reported that ⅓ acknowledged some income from representing tort claimants, but for only 20% was it the most important field of practice. Widiss and Bovbjerk, 'No-fault in Massachusetts: Its Impact on Courts and Lawyers', 59 *ABAJ* 487 (1973).

of the poor who predominantly constitute the class of tort victims and therefore of potential plaintiffs. The upper class lawyer would become involved on behalf of his business clients in tort as in other litigation; in more modern times, largely in the wake of liability insurance, tort defence work has become a specialty of (private practice) insurance lawyers, though not exclusively.

What divides the two groups today is principally a function of the economics of the profession. Plaintiffs' attorneys are dedicated to the contingent fee, while the defence bar is paid on an hourly rate. The distinction is beset with deep ethical and social prejudices, creating a gulf that is rarely crossed even for an occasional change of roles.[3] This polarization is aggravated by its institutionalization—the American Trial Lawyers' Association (ATLA) representing the plaintiffs' bar, the Association of Insurance Counsel, and some other minor organizations representing the defence bar. The former, especially, has contributed a great deal to the sense of identity and advancement of common purposes to which its members are publicly committed. Their rhetoric as well as their political activities emphasize confrontation. This attitude is not only replicated by the other side, but aggravated by attacks on their integrity and ethics. The relationship of the two sides, none the less, is symbiotic since both profit from the opposition of the other. Just as there would be no liability insurance in the absence of liability, so an aggressive plaintiffs' bar feeds the practice of insurance lawyers. Conversely, plaintiffs' lawyers would have to content themselves with much smaller fees if they did not have to earn them the hard way. Both therefore share an abiding faith in the adversarial tort system—their bread and butter.[4]

Besides these forensic gladiators, insurance adjusters play, if anything, an even more important role than defence attorneys. Most tort defendants worth suing are represented by liability insurers whose adjusters conduct the settlement negotiations, which in all but a small percentage of cases

[3] Defence lawyers may be on both sides in indemnity litigation and the like.

[4] One is reminded of the lawyer's prayer: 'Dear Lord, let there be strife among the people, so that your loyal servant may prosper.'

terminate the affair. Settlement rather than trial is the principal rite of tort litigation.

In the following pages, I therefore propose to deal in turn with plaintiffs' attorneys, defence attorneys, and insurance adjusters. A seperate chapter will address the operation and effect of the contingent fee on tort liability.

1. THE PLAINTIFFS' BAR: ATLA

1.1 In the beginning

Early tort lawyers 'were far from the heroic figures that one imagines in the role of "keeper of the poor man's key to the courthouse"'.[5] By the middle of the nineteenth century, burgeoning transport and manufacturing were taking a spectacular toll in injury and death especially from the working population. The tort system responded very inadequately to this challenge. Whatever their motivation,[6] the courts offered little encouragement to tort claims; indeed they developed defences which largely stultified attempts to bring negligent employers and other defendants to book.[7] Beyond that, access to justice was hindered both by the social and economic handicaps of the victims and by unconscionable tactics on the part of the defence. Strong-arm methods, perjured testimony, and corruption were widely practised by both sides, each justifying itself as acting merely in self-defence or retaliation.[8] No wonder that personal injury litigation came to be scorned as 'dirty business' by respectable members of the profession and abandoned to a subclass of lawyers unable to eke out a living in a more rewarding practice. Élite lawyers from

---

[5] Speiser, *Lawsuit* 122 (1980).

[6] One theory has it that the courts deliberately structured tort law to subsidize emerging industries at the cost of the working classes. e.g. Horwitz, *The Transformation of American Law 1780–1860* (1977); Friedman, *A History of American Law* (1973). But see Schwartz, 'Tort Law and the Economy in Nineteenth-Century America: A Reinterpretation', 90 *Yale L. J.* 1717 (1981).

[7] See, e.g., Friedman and Ladinsky, 'Social Change and the Law of Industrial Accidents', 67 *Colum. L. Rev.* 50 (1967).

[8] Speiser, *supra*, at 139–41.

Abraham Lincoln to Clarence Darrow acted for the rail-roads, not against them. The latter business was left to lower-class lawyers, handicapped by immigrant origins of non-Anglo-Saxon and non-Protestant stock, graduates of night schools, to whom the doors of respectable law firms were shut. Within the legal profession their pariah status was emphasized by disdain for their contingent fees to which they had to resort on behalf of generally impecunious clients.

The big firms were insulated from this 'intellectual slum', as David Riesman has called it, where 'a largely ethnic bar carries on the Anglo-Saxon rites of trial by jury and "contaminates" the legal ideal with demagogic practice'.[9] Another irony was that these despised practitioners of the ancient arts were, and for that matter still are,[10] largely solo practitioners, retaining the ideal of America's rural past, while the Anglo-Saxon scions of the corporate world were concentrated in ever larger city firms.[11] But instead of hiding their embarrassment, the establishment, long identified with the American Bar Association (ABA), went on the offensive. The campaign of denigration against plaintiffs' lawyers, in-termittently waged by élitist members of bar associations, focused both on contingent fees and on legal ethics. The least worthy of its motivations, it has been suggested, was that it concealed class and ethnic hostility.[12] In any event, the censure seemed the more hypocritical as corporate lawyers, no less than their brethren in the trenches, had all become businessmen—a transition from learned profession to business which had

[9] Riesman, 'Law and Sociology: Recruitment, Training and Col-leagueship', 9 *Stan. L. Rev.* 643, 665 (1957).

[10] Carlin in his two studies *Lawyers on Their Own* (1962) and *Lawyers' Ethics: A Survey of the New York City Bar* (1966) has given an insightful account of the solo practitioner's plight in Chicago and New York continuing in modern times. See also Ladinsky, 'The Impact of Social Backgrounds of Lawyers on Law Practice and the Law', 16 *J. Leg. Ed.* 127 (1963) on the continuing coincidence of solo practice with ethnic and socio-economic factors.

[11] The withdrawal of the élite lawyer from the courtroom to the role of counsellor to business interests is one of the themes of Bloomfield, *American Lawyers in a Changing Society, 1776–1876* (1976).

[12] Auerbach, *Unequal Justice (Lawyers and Social Change in Modern America)*, ch. 2 (1976).

started after the Civil War.[13] Moreover, apart from supporting a marginal role for Legal Aid Societies,[14] the legal establishment contributed little to relieve the plight of accident victims and thus implicitly ratified the status quo, including the practices of the segregated plaintiffs' bar. After all, is not the contingent fee a reminder of the inadequacy of legal services, for which the whole profession bears responsibility?

Not until well into the twentieth century did the situation begin to improve. The passage in 1908 and 1920 respectively of Congressional legislation on railroad and maritime accidents, and of Workmen's Compensation Acts in the States beginning with the New York statute in 1910, brought substantial improvement for industrial victims.[15] The first two in particular, by facilitating tort recovery, gave rise to the first group of American tort lawyers who could face their adversaries with any semblance of equality.[16] Their ability to specialize and reap substantial rewards from their contingency fees enabled them to develop the necessary expertise and professionalism to stand comparison with traditionally more prestigious branches of the bar. Although motor-car accidents grew to mammoth proportions in the 1920s, their contribution to the success of trial lawyers was nowhere near as spectacular. Workmen's compensation, which in the United States replaced tort remedies against the employer, in turn gave rise to a specialized practice which, however, was never as lucrative as tort litigation because fees were limited to a comparatively small statutory percentage of modest benefits.

The increasing practice of referrals also contributed to the greater success of tort claims, not however without en-

[13] The disintegration of the old order towards the mid-century, attributed in part to 'Jacksonian democracy', is traced by Chroust, *The Rise of the Legal Profession in America*, vol. ii (1965).

[14] Ibid. 53–62.

[15] This is reflected in the steep rise of personal injury litigation. Tort cases before state Supreme Courts rose from 5.7% in 1870–80 to 16.4% in 1905–35 and 22.3% in 1940–70. During 1885–1920 ⅓ of all tort suits arose out of railroad and streetcar accidents. Thereafter they were overtaken by automobile accidents, which rose from 2.1% in 1915–20 to 7.6% in 1940–60. Kagan, Cartwright, Friedman, and Wheeler, 'The Business of State Supreme Courts, 1890–1970', 30 *Stan. L. Rev.* 121, 142–5 (1977).

[16] Speiser, *supra*, at 145.

countering a chorus of condemnation as another form of ambulance chasing. Two official inquiries into the negligence practice in New York, the first in 1928, the second in 1955, mark the notable improvement in performance and ethics during the intervening period.[17] The great leap forward in the fortunes of the trial bar coincided with the founding of its own professional organization, now known as the American Trial Lawyers Association (ATLA). It marked the transition from the old order to the new—the order of the entrepreneur.

1.2   ATLA and its ideology

The Association was modestly founded in 1946 in Portland, Oregon, under the name of National Association of Claimants' Counsel Attorneys (NACCA) by a group of workers' compensation attorneys, later joined by lawyers representing railway workers, admiralty, and aviation claims. Membership rose rapidly from the original 25, reaching 2,000 by 1951, 7,000 by 1960, and swelling to its present total of over 65,000.[18]

ATLA's national organization has been reinforced by state affiliates whose activities mirror the parent's at the local level. Most influential is the California Trial Lawyers' Association (CTLA) with a current membership of 5,400, whose political and educational activities can be traced in its monthly journal, *CTLA Forum*.

From the start, the history of the organization became marked by acrimonious confrontation with the insurance industry, one side proclaiming the need for reforming the tort system, particularly the jury and lawyers' contingent fees, the other castigating the tactics and practices of insurance companies. These altercations reached a larger public forum in 1952 with a series of articles in the *Reader's Digest* attacking

[17] The 1928 Wasservogel Report in the *Judicial Investigation of 'Ambulance Chasing' in New York* (1928) was published in 14 *Mass. L. Q.* 1–32 (1928). The Report of Special Referee Wasservogel, *Regarding Proposed Rule to Limit Compensation of Plaintiffs' Attorneys in Personal Injury and Wrongful Death Actions* (1956), is discussed in Note, 'Lawyer's Tightrope: Use and Abuse of Fees', 44 *Corn. L. Q.* 683 (1956).

[18] Of some 700,000 American lawyers 300,000 are members of the ABA, a voluntary organization. Only a few states, like California, require membership in a state bar organization.

NACCA lawyers, presumably at the behest of the insurers. ATLA ideology is formed by an intensely individualistic perspective of private citizens wronged by corporate forces: big business, airlines, railroads, the maritime industry, and the insurance machine. To this may be added, since the unleashing of the civil rights action in 1961,[19] oppressive government officialdom. 'Secretly, these men are suing society.'[20]

A central feature of this ideology is the veneration of jury trial as the birthright of every American, indeed the incarnation of Americanism. Juries represent the common man and can therefore be trusted to express their identification with and compassion for plaintiffs, their peers. Cynics might add that juries are more manipulable by appeals to prejudice and passion and, in returning big awards, help not only the victim but also his lawyer's fees. The ATLA lawyer's stake in the jury is matched by intense attention to the process of jury selection (*voir dire*) and concern about opposition moves to curtail the province of the jury.

There is a distinctly old-fashioned sound of evangelical fervour to ATLA's rhetoric. Added to their idealization of the grass roots link of the jury is their high moral view of tort law itself. Tort liability has the purpose of wreaking retribution for a wrong done to an innocent victim. The wrongdoer must account as an individual for the injury inflicted on another. Liability thereby serves both to correct the injustice to the victim and as a warning to others. There may be other auxiliary sanctions against unacceptable conduct, but experience shows that none equals tort liability in its effectiveness. Tort law provides the necessary incentives for public champions to root out practices dangerous to the community and bring the perpetrator to public judgment. Indeed, nothing is so satisfying to that cause as the award of heavy punitive damages. Hence, modern notions about loss distribution, economic efficiency, and the like are viewed with suspicion as projecting the role of tort in wider ideological terms with undesirable implications

---

[19] *Monroe* v. *Pape*, 365 US 167 (1961). Since then, civil rights filings have mushroomed from 296 in 1961 to over 12,000 in 1985.

[20] Mayer, *The Lawyers* 266 (1966), citing an unidentified 'observer from the law schools'.

for the cost burden to society at large, rather than as a simple issue between victim and oppressor. The irony of the ATLA posture is that plaintiffs' attorneys are interested in pursuing only defendants with deep pockets, i.e. those who either carry hefty insurance or are otherwise able to pass on the cost of adverse verdicts to a wider segment of the market—in other words, defendants who often do not match the stereotypical wrongdoer who must answer with a pound of his own flesh.

The trial lawyers' contingency fee constitutes their most sensitive nerve. Instead of revealing a mercenary motive, it is portrayed as a device principally in the client's interest. It is 'the poor man's key to the court-house'. But it is also the most controversial feature of the American tort process. Its central role deserves a separate chapter, following next.

For this and other reasons, the ATLA rhetoric is viewed with considerable scepticism by many. For every value touted also happens to be one that increases their business and income. They serve low status clients because they could not find any bigger. They drum up business by shamelessly soliciting in the media[21] and chasing ambulances,[22] not as Good Samaritans but because their living depends on it. They advocate jury trials because they bring higher awards, greater percentages in contingency fees, and more chances of success.

---

[21] The traditional prohibition against soliciting was lifted when the US Supreme Court nullified the Virginia statute as unconstitutional, in *NAACP* v. *Button*, 371 US 415 (1963). The US Supreme Court upheld the ban against person-to-person solicitation (*Ohralik* v. *Ohio St. Bar Assn.* 436 US 447 (1978)), but reaffirmed its protection of general advertising (*Bates and O'Steen* v. *State Bar of Arizona*, 433 US 350 (1977)), including illustrations (*Zauderer* v. *Office of Disciplinary Counsel*, 471 US 626 (1985)). By 1984, 13% of US attorneys were advertising through print or broadcast media, according to a poll conducted by the ABA. Statistics compiled by the Television Bureau of Advertising show that lawyers spent $28m. on TV advertising that year. Direct mail solicitation of potential asbestos claimants among certain occupational groups was reported in *Wall St. J.*, 18 Feb. 1987.
Perhaps the most shocking is the collection of clients by advertising and their wholesale referral (for payment) to other lawyers: *Wall St. J.*, 3 June 1986, reporting on a Boston attorney with an 'inventory' of 1,700 cases.

[22] See Carlin, *supra*, n. 10. Note, 'Legal Ethics: Ambulance Chasing', 30 *NYU L. Rev.* 182 (1955); Reichstein, 'Ambulance Chasing: A Case Study of Deviation and Control Within the Legal Profession', 13 *Social Problems* 3 (1965); Monaghan, 'The Liability Claims Racket', 3 *L. & C. P.* 491 (1936).

They favour state rather than federal control because their cronies are judges, councilmen, labour union bosses, and legislators. Being closer to the grass roots, they can indeed participate in ('buy, manipulate, influence') the political and societal processes to their advantage.

1.3 Self-image

A dominant and persistent objective of the organization and its members has been to improve their public image, not least within the purview of the legal profession itself. One way was to emphasize the informational and educational activities of the organization, and thereby assist in raising the professional competence of its members and vie with the prestigious activities of the American Bar Association and the American Law Institute (ALI-ABA).

From its very beginning in 1946 ATLA has held an annual convention and in 1948 commenced publication of the *NACCA Law Journal*, which, after several name changes, is now a monthly called *Trial*.[23] From 1953 to 1955 Roscoe Pound, the famous former Dean of the Harvard Law School, became its editor, thereby considerably enhancing the respectability of the organization itself. In 1957 the *NACCA Newsletter* (now the monthly *ATLA Law Reporter*) was added to the *NACCA Law Journal*, followed in 1960 by the *Public Information and Educational Bulletin* as well as a national Speakers Bureau and a department of public information and education. One of the most useful and successful organs, the *Exchange*, is a research service containing more than 4,000 files on products liability, automobile, and other negligence cases, with access to technical information, expert witnesses, and numerous data bases.[24]

The expanding orientation and self-image of the organization was reflected in its change of names. In 1960 it became the National Association of Claimants' Counsel of America, in 1964 the American Trial Lawyers' Association (ATLA).

[23] *ATLA Journal* (1964–78), *Trial* (1978).
[24] *ATLA Advocate*, vol. 12, no. 7, p. 1. An early example of its effectiveness was its role in the MER/29 litigation: see Rheingold, 'The MER/29 Story', 56 *Calif. L. Rev.* 116, 122 (1968).

This was accompanied by a deliberate policy to shed the ambulance-chaser image and to disavow the more histrionic and abrasive behaviour of some of its members, notably the bad boy of the trial bar, the self-proclaimed King of Torts, Melvin Belli of San Francisco. It had become Belli's practice each year in conjunction with the ATLA summer convention to give a series of upbeat seminars on successful trial techniques, especially so-called 'demonstrative evidence' which he claimed as an invention of his own.[25] His forensic success in presenting maimed clients to the jury coined a household word (a 'basket case');[26] he was also the first to develop a videotape for courtroom presentation of one day in the life of a paraplegic client. But his public posturing and self-advertising, including a notorious television show and bringing the infamous mobster Mickey Cohen to a national convention, had become so controversial that he was barred from the 1965 convention, although not banished for ever. Twenty years later he was still giving trouble and embarrassing the public image of the trial lawyers by his scandalous client-chasing in the wake of the Bhopal tragedy.[27]

Self-interest as well as self-image are involved in ATLA's recent opposition to the ABA's Revised Canons of Ethics and its counterproposals. Beneath the merits of the specific issues at stake here is the distribution of power in the legal profession. Higher social status and wealth are widely believed to control the ABA, which defines the ethical standards of the entire profession. Not surprisingly their precepts were often perceived to be ill-adapted to the realities of small-time marginal practices, especially its canons regarding solicitation, referrals, fee-splitting, and the like. Personal injury lawyers have long complained that attorneys in large firms practising corporate law do not have to stoop to soliciting clients; clients come to

---

[25] See Belli, *Modern Trials* (1954).

[26] Belli, *Ready for the Plaintiff*, ch. 24 (1956).

[27] Without mentioning names, the ATLA board of governors resolved to condemn 'the conduct of attorneys or their representatives who go, uninvited, to the scene of a disaster, set up temporary quarters and advertise for prospective clients . . . ': *Trial*, Mar. 1986, p. 5. A year earlier the CTLA board had likewise condemned 'a most distasteful level of "hucksterism" employed' by uninvited American lawyers: *CTLA Forum*, Feb. 1985, p. 2. Once more Habush (ATLA Pres.), *Trial*, Apr. 1987, p. 7, calling for disciplinary action.

them as part of an invisible and social upper-class network so that they can afford to maintain a dignified and aloof posture, untainted by 'trade'. In contrast, the proscribed practices are often essential for the survival of the personal injury lawyer. Although violations may go without disciplinary sanctions, they make it difficult to maintain membership, let alone attain positions of power and prestige, within respectable bar associations. Their practices are forever branded as unethical and their status as that of pariahs.[28]

Hence when the trial lawyers proposed their own alternative to the Kutak Commission's new Code of Ethics, it marked not only their disagreement with specific points of practice, but a resounding protest against their positions as second-class citizens of the bar. This intervention was well timed to coincide with the remarkable shift in the structure and fortunes of the personal injury bar. Ironically, this development was not unconnected with the success of automobile no-fault insurance which ATLA had opposed for so long. Traffic accidents used to constitute the mainstay of the small-gauge practitioner. The contraction of this business and its replacement by products and medical liability, among others,[29] in the 1970s accelerated the trend of new ambitious entrepreneurs superseding the marginal practitioners of yester-year in the ranks of the personal injury bar and ATLA's corridors of power. Products liability litigation, like technology, requires an educated and sophisticated body of lawyers. Graduates from prestige law schools, with prior engineering, medical, or economics degrees, are now attracted into the profession by its highly publicized

[28] The economics of the profession, rooted in the contingent fee, which drive these ethically suspect practices are succinctly described by Rosenthal, *Lawyer and Client: Who's in Charge?* ch. 4 (1974, 1977). Further *infra*, ch. 6.

[29] The Rand Study by its Institute of Civil Justice reported a substantial deviation in the relative size of awards between automobile cases on the one hand and products liability and work injury on the other. While the former decreased in the 1960s and 1970s (median from $8,000 to $5,000), the latter increased substantially in the last decade (to $82,000). The author surmised that this trend could, *inter alia*, reflect specialization among plaintiffs' lawyers. Peterson, *Compensation of Injuries. Civil Jury Verdicts in Cook County* 55–6 (1984). The same trend was observed in the San Francisco study by Shanley and Peterson, *Comparative Justice. Civil Jury Verdicts in San Francisco and Cook Counties, 1959–1980* (1983).

rewards. Today's leaders are 'fat cats', flamboyant still but wealthy beyond imagination and, for both reasons, admired by the public and their own. Their Olympic circle is the Million Dollar Club of members who have recovered verdicts of that amount and more, and whose contingent fees in consequence often far exceed the income of senior partners in corporate practices. These are the lawyers who successfully battle industrial giants, motor-car manufacturers, pharmaceutical companies, and the airline industry.

The complexity of much of this litigation has entailed other adjustments. Because the costs are very high and have to be borne at least initially by the attorney himself, concentration into the hands of the successful has intensified.[30] And because most plaintiffs' firms employ fewer than twenty lawyers, in contrast to the larger defence firms, the former have been driven to make up for the disadvantage in resources by teaming up in mass litigation, contributing to common defence funds and fee-sharing agreements. In class actions and other consolidated trials a lead 'team' of the most prominent and expert typically take the helm.

The underclass of 'unethical' and disreputable attorneys, far from still being the model or average, is being disowned, and occasionally embarrassing ambulance-chasing in mass accidents is publicly censured.[31] Even if personal injury work still ranks lowest in peer estimation,[32] at all events the trial bar has become powerful on its own terms even within the councils of the once contemptuous ABA.[33]

[30] The degree of concentration in larger cities was already evident in the late 1950s, as revealed in the New York study by Franklin, Chanin, and Mark, 'Accidents, Money, and the Law: A Study of the Economics of Personal Injury Litigation', 61 *Colum. L. Rev.* 1, 47 (1961). Their estimate was that of about 15,000 plaintiffs' lawyers, 300 were specialists (averaging more than 100 recoveries p.a.) who handled about 20,000 personal injury cases annually.   [31] *Supra*, n. 27.

[32] See Heinz and Laumann, *Chicago Lawyers* 107-8 (1982) ('dirty' or 'unsavory').

[33] e.g., the ABA sided with ATLA in rejecting the current proposals for medical liability reform (limiting awards, contingent fees, etc.): *New York Times*, 12 Feb. 1986, p. 21. Cf. 'an influential portion of the Bar . . . organized as never before to promote ever-increasing recoveries for the most intangible and elusive injuries': *Battalla* v. *State of New York*, 10 NY 2d 237, 243, 176 NE 2d 729, 732, 219 NYS 2d 34, 38-9 (1961) (Van Voorhis, J.).

1.4 Political activities

In the first decade of its existence, ATLA's orientation was primarily focused on the courtroom. Since then, its activities have shifted increasingly to the political arena. Until fairly recently, these centred principally on state legislatures rather than on the US Congress. This preference reflected, in the first instance, the state-law identity of most personal injury law. The nature of local politics, particularly in the big cities with their overwhelming Democratic majorities, also assured reliable allies among legislators for ATLA's causes. Personal injury lawyers had traditionally cultivated close relations with influence pedlars in city halls and county offices,[34] so that their move on the state capitol was but a natural extension of their local, folksy approach to mobilizing support in the corridors of power.

Here their concern has been with two principal goals: one to influence the selection of judges, the other to influence legislation. In jurisdictions where judges are elected, the selection of candidates can be influenced through party channels at either the county or the state level (trial and appellate judges respectively). Where judges are appointed, as in California, access to the governor's office, directly or through political allies, becomes rewarding. The value of sympathetic, pro-plaintiff judges has been keenly recognized by the plaintiffs' bar. Trial judges play an active and dominant role in forcing settlements; they influence the outcome of litigation by rulings on the evidence and, most important, by controlling whether the case goes to the jury and under what instructions. Appeal judges exercise a great deal of discretion in supervision of the trial process, and the Supreme Court usually has the last word on what the law is. At all these levels, it pays handsomely to have friends in court. The remarkable pro-plaintiff trend of tort law in the post-World War II period could not have been accomplished without the active support or connivance of a sympathetic judiciary. The election or appointment of predominantly liberal judges in most of the populous states has been the harvest of a crop

---

[34] See Carlin, *supra*, n. 10.

assiduously cultivated by ATLA and its political allies. In California, for example, Governor Jerry Brown (1973–81) in his mission to change the face of the judiciary by wide-scale preferment of women, minorities, and lawyers outside the mainstream, reportedly consulted CTLA on all appointments.[35] As a result, more than half of California's judges, those appointed by him, are basically sympathetic to the cause. So especially was the Supreme Court which, as noted in earlier chapters,[36] consistently demonstrated a pro-plaintiff bias in tort appeals. Its perceived concomitant pro-defendant bias in criminal cases, especially those involving death sentences, eventually caused a popular backlash which resulted in the defeat at the polls of the Chief Justice, Rose Bird, and two of her colleagues. The CTLA became the principal standard-bearer and financial supporter of the embattled threesome, mindful of its gratitude for past and, hopefully, future favours.

1.5 Legislative battles

In the legislative arena, the plaintiffs' bar has been mostly engaged in fending off moves to undo the effect of favourable rulings by the courts or to impose restrictions on tort remedies. Two major issues in particular have been at the top of ATLA's and its local affiliates' agenda: no-fault automobile plans and the so-called insurance crisis.

In the 1960s, some of the insurance industry had found an ally in the American Bar Association for proposals, eventually carried as far as the US Congress, to abolish jury trial in automobile insurance cases. As insurance rates were steadily climbing to allegedly intolerable levels, blame was falling on the jury system for the high awards, the high cost of litigation and the court congestion in many jurisdictions. The ABA proposed revision, even limitation of contingent fees, while sectors of the insurance industry were hatching no-fault automobile insurance plans. ATLA bitterly opposed those containing restrictions on tort recovery, but for obvious reasons

---

[35] Bancroft, 'All Deukmejian's Judges', *Calif. L. J.*, Jan. 1986, p. 49.
[36] *Supra*, chs. 2 and 3.

supported so called 'add-on' plans which assured minimal benefits to victims without impairing their common-law rights. Indeed, it was the legislative battles on this issue that drew ATLA's attention to the importance of cultivating the legislative scene. Their widespread success in defeating insurance no-fault bills or in emasculating them into harmless add-on plans[37] raised their public profile and increased confidence in their ability to play a dominant role not only in the courtroom but equally in the legislative chamber.

The medical crisis of the mid-1970s was precipitated by the threatened withdrawal of liability insurers from the field of medical liability because of heavy losses allegedly due to an explosion of jury awards, unexpected both in numbers and in size. Hospital doctors in some cities went on strike, refusing to attend patients without insurance cover. To the extent that insurance continued to be available, premiums sky-rocketed, driving some physicians into retirement. In order to avert this crisis, proposals emerged for setting up state insurance pools and limiting the cost of liability in various ways, such as limits on damages for non-pecuniary loss and contingent fees, awards in the form of periodical payments, and the reduction of awards for collateral benefits such as prepaid medical insurance. All these measures incurred the bitter opposition of ATLA, which blamed the whole crisis on the insurance industry. The real reasons, it argued, were the precipitous fall in the value of insurance investments due to the recession of 1973-5 and the failure to increase rates in a blind search for business during a credit squeeze.[38] As for the rise in insurance rates, its cause was attributable to medical incompetence and the conspiracy of the profession to protect the charlatans in its own ranks— two evils which it was the mission of tort lawyers to root out. Exemplary damages were an incentive to medical safety and professional supervision as well as a monetary calculus for moral injury ('fault'). Legal fees were admittedly high but necessitated by the technical nature of the issues and the cost

---

[37] See *infra*, nn. 95-104.
[38] Current campaign literature by CTLA is illustrated by cut-outs in 1985 *CTLA Forum* 183-201.

of expert witnesses. The stakes were larger in the event of victory, but so too were the chances of defeat.[39]

The ATLA litany, however, was not proof against the profound public concern on this issue and limiting legislation passed on to the statute book in many states. Nor in this instance did ATLA succeed in successfully challenging the constitutionality of these reforms.[40]

ATLA was also forced into a defensive posture in resisting industry-inspired legislation designed to roll back some of the successes it helped to score on behalf of plaintiffs in the judicial development of strict liability for defective products. Inspired by a Model Products Liability Act, under the auspices of the federal Department of Commerce, legislation was launched in many states to introduce curbs on punitive damages, modify liability for design defects, and introduce a statute of repose which would extinguish all claims after a certain period (say ten years) regardless of whether the injury, or exposure to it, was discoverable. Despite stiff opposition from ATLA, one or more (though rarely all) these measures were eventually enacted in many states, especially in the period of 1985-7.[41]

ATLA's consolidation as a political force gradually extended also to the national scene. In 1967 it appeared before the House of Representatives on jury selection laws in train with its growing involvement in criminal defence issues; in that and the following years it successfully opposed federal no-fault automobile legislation in the US Senate. In 1968 ATLA members became advisers to the National Commission on Consumer Safety, and in 1969 testified in opposition to President Nixon's nominations of Judges Hainsworth and Carswell to the US Supreme Court.

In the 1970s new issues surfaced: Legal Services and Products Liability curbs. In 1971 ATLA, in company with bar associations, supported increased federal funding, indeed a 'Magna Charta', guaranteeing to the poor the right to

[39] According to the Chicago jury study, the success rate of malpractice actions 1959-79 was 29%, compared with 50-2% overall. See Peterson, *supra*, at 38-44; Shanley and Peterson, *supra*, at 50-3 for comparison with San Francisco.
[40] See *supra*, ch. 3.
[41] See *supra*, ch. 1, n.85.

affordable legal services. In 1973 it supported, in opposition
to the ABA, an amendment to the Taft–Hartley Act providing
for union prepaid legal services plans. When questioned on
its motivations, the ATLA president pointed to its own *pro
bono* plan whereby each member, then some 30,000, would
undertake one *pro bono* case a year.

The most important Congressional issue for ATLA in the
last decade has been to resist efforts to impose curbs on
products liability through federal legislation. As already noted,
the campaign by industry to effect such reforms in the States,
their natural forum, met only with indifferent success in face
of the stiff opposition by ATLA and its allies; in order to be
effective, given a national market for most products, adoption
had to be near universal. The reformers' attention therefore
shifted to the national forum, in the hope also of gaining a
more industry-sympathetic hearing in Washington. ATLA has
been opposing industry bills with vastly increased resources
since the early 1980s.[42]

1.6 PACs

ATLA's legislative successes have been the reward of assiduous
cultivation of members of the legislatures. Many of these,
especially members of the strategically important judiciary
committees, are trial lawyers themselves or belong to firms
engaged in that line of practice. Because of their general
ideological orientation, previously described, most if not all
are supporters of the Democratic party, which, in the more
populous states with big urban populations, generally constitutes the majority party. Key legislators, like the Speaker
and committee chairmen who control the legislative agenda,
therefore tend to be sympathetic to ATLA's concerns.

Not content with these natural allies, ATLA and its local
affiliates have sought to increase their hold by giving financial
support to old friends and new. As early as 1971 it set up a
Department of Federal and State Relations and assigned to

[42] It now has a Washington staff of about 100 and retained premier
lobbyists, Democrat T. Boggs and Republican W. Timmons: *Wall St. J.*
9 Apr. 1986, p. 62. The principal bills are carried by Sens. Kasten and
Danforth.

it 25 per cent of its total budget. Like many other trade and ideological organizations, ATLA also organized a Political Action Committee (PAC) under the name of the Attorneys' Congressional Campaign Trust. In 1980, it was estimated that ATLA, then having a membership of 43,000, contributed between $300,000 and $400,000 in Congressional campaigns to 30–5 Senators and more than 200 Representatives. Between 1981 and 1982, with approximately 50,000 members, ATLA's PAC collected $889,000 and contributed $449,000 to 332 Congressmen.[43] These amounts were, proportionate to their membership, very much larger than even those disbursed by the formidable American Medical Association (AMA).[44] Most of the larger contributions (above $1,000) went to members of congressional committees that control legislation of potential concern to ATLA, such as the Senate Committee on Commerce, Science, and Transportation (products liability), the House Committee on Energy (products liability), the House and Senate Judiciary Committees (diversity jurisdiction of federal courts), and the House Rules Committee which controls the flow of legislation to the floor.

This pattern is duplicated at the state level by ATLA's affiliates like the California Trial Lawyers Association. In 1985 it was reported that CTLA had contributed $91,000 to current members of the Assembly Judiciary Committee during the previous two years, more than ten times the amount paid by the Farm Bureau. The committee chairman, a Democrat from Oakland, received nearly $20,000 in contributions between 1982 and 1984. CTLA support has not been confined to members of the Judiciary Committee. The Assembly Speaker, who controls membership of that committee, received $124,753 in 1983–4. Between 1975 and 1984 CTLA's PAC distributed $1.42m. to state legislative candidates, the sixth highest contributor;[45] in 1986 alone almost $700,000.[46] This largesse

---

[43] Federal Election Commission (D Index).

[44] $1,644,795 for four times the membership (270,000). In 1983–4 ATLA ranked 31st in receipts (45th in expenditures) among all PACs, 8th (and 10th respectively) among trade associations.

[45] Common Cause, *Report, San Francisco Chronicle*, 23 May 1985.

[46] An average of almost $5,000 per legislator. Additionally, CTL FedPAC distributed more than $100,000.

is forthcoming not only in election years when the pretence is more credible that it is merely a contribution to election expenses which have sky-rocketed in the television era. It has flowed even in off-years when influence-peddling is its most obvious purpose. Significantly, the top off-year contributor in 1983 was the CTLA, which gave $237,000 to seventy legislative incumbents.[47]

As previously mentioned, the defence of Chief Justice Bird in her confirmation battle for a second term was principally financed by CTLA and individual members. Her notorious partisanship for tort plaintiffs had earned her the accolade of CTLA's 'friend on the court'.[48] This favour was now being repaid by solid financial support in aid of 'judicial independence'. If campaign contributions to political candidates are difficult enough to justify as an ethical practice, those in support of judges are even more questionable. At the very least they illustrate and aggravate the hazard of politicizing the role of courts.[49]

1.7   *Amicus* briefs

For promoting its causes in the courtroom, ATLA has seen one of its principal functions to be educating its members through its periodicals and through seminars and lectures in connection with and between annual conventions. Additional resources are the data bank previously mentioned, information centres on current mass litigation as well as model briefs and associated materials.

Even more remarkable is ATLA's involvement, and its local affiliates, in the role of *amicus curiae* in litigation of concern to its membership. In order to place this activity in perspective, a word must first be said about the evolving role of the *amicus* brief generally in American litigation.

[47] *The New Gold Rush: Financing California's Legislative Campaigns* (1985 Report and Recommendations of the California Commission on Campaign Financing). It recommended prohibition of all off-year contributions and voluntary financial limits on others.

[48] Lest it escape the reader, this word-play on 'friend *of* the court' evokes CTLA's *amicus curiae* activities.

[49] See Lowenstein, 'Political Bribery and the Intermediate Theory of Politics', 32 *UCLA L. Rev.* 784 (1985).

Its ancient origin, going back to Roman law, and its long history in the common law are apt to obscure the changed function and importance of the device in modern American law.[50] The English common law did not favour intervention in litigation by third parties and looked upon the *amicus curiae* as true to its name, a friend of the court. The Year Books record numerous incidents of bystanders drawing the court's attention to relevant statutes or facts to avoid error. A forerunner of its more modern role was a Member of Parliament who informed the court about the intent of Parliament in passing the legislation.[51] At a fairly early stage, also, we encounter *amici* uncovering collusive suits prejudicial to an officially unrepresented party. This role marked a transformation from a disinterested friend of the court to one committed to a specific cause.

In America the transition was hastened by the circumstances of the developing practice of the US Supreme Court in the course of the nineteenth century. The development was an inevitable response to the wider impact, beyond the immediate parties, of litigation under a federal constitution. State rights and other public interests were often implicated in the issues presented by private parties but inadequately represented by them. By the 1880s the court began to grant leave directly to state counsel to vindicate state rights. At the same time it also expanded the right of participation of private litigants, such as those with similar cases pending before lower courts and others who might be directly affected by the decision. Increasingly, those interests were represented by organizations, although well into the twentieth century the fiction of the essentially professional relation of the *amicus* to the court was maintained by not explicitly attributing authorship of the brief to the organization itself. This could not, however, obscure the reality of the transition: 'the institution of the

---

[50] See Krislov, 'The *Amicus Curiae* Brief: From Friendship to Advocacy', 72 *Yale L. J.* 694 (1963). This article, like all others, focuses on US Supreme Court practice: see also Harper and Etherington, 'Lobbyists Before the Court', 101 *U. Pa. L. Rev.* 1172 (1953) ('probably the path-finding article on this topic'); O'Connor and Epstein, '*Amicus Curiae* Participation', 16 *L. & Soc. Rev.* 311 (1982).

[51] *Horton* v. *Ruesby*, Comb. 33 (1686), cited by Krislov, *supra*, at 695.

amicus curiae brief ha[d] moved from neutrality to partisanship, from friendship to advocacy.'[52]

This changed role of the *amicus* brief also reflected an ongoing transformation in the modes of 'special interest' activities. Increasing bureaucratization brought with it greater ability to mobilize resources in promoting 'political' causes in the legislature, administration, and also in the courts. The example set by public agencies was duly followed by private organizations, first those of a business or commercial nature, later joined by representatives of previously voiceless minorities in the cause of civil rights, like the American Civil Liberties Union (ACLU), the National Association for the Advancement of Colored People (NAACP), the American Jewish Congress, and others. This has helped somewhat to democratize the process of judicial adjudication by giving the court access to a wider spectrum of views and thereby mitigating the limitations inherent in the judicial process compared with the more open forum for public debate and political interplay in legislatures. By the same token, it has helped to meet the widely felt demand for popular participation in the decision-making process as a desirable postulate of democratic government. The central role of the US Supreme Court in defining social agenda makes it all the more desirable that, despite its tenuous democratic credentials, it should not be perceived as isolated from the currents of a pluralistic society.[53]

Awareness of law as a process of social choice and policy-making is not confined to the role of the US Supreme Court or adjudication generally of constitutional issues. The message of the Realist school of jurisprudence, which tore away the mask of judicial neutrality for the generation of the 1930s,[54] has since been all too amply reinforced by the bolder activism of courts in general. No doubt the model of federal courts seized of constitutional issues contributed to this changing perception of the judicial role; another reason is the widely shared view that courts must play a major role in effecting

---

[52] Krislov, *supra*, at 704.

[53] Hakman, 'Lobbying the Supreme Court: An Appraisal of "Political Science Folklore" ', 35 *Fordh. L. Rev.* 15 (1966).

[54] See especially Frank, *Law and the Modern Mind* (1930); id., *Courts on Trial* (1949).

social change since legislatures are unable to do so on their own. As issues presented to the courts in private litigation in consequence became more controversial and 'public' in their wider potential impact on society, the scene has become increasingly similar to that in constitutional litigation. The intensity and plurality of affected interests therefore called for similar opportunities of access to the decision-making process. Not the least important of these is the *amicus* brief.[55]

The *amicus* brief also helps to protect courts against less desirable ways of exerting external influence. State court judges are rather more vulnerable to such importunities in view of their involvement, past if not present, in the political life of their State. The *amicus* brief has become a respectable and discreet channel for such communications.[56]

ATLA and its local affiliates have engaged in sustained programmes to intervene in tort appellate litigation to support their professional causes, occasionally as direct litigants,[57] more often as *amici curiae*. Although not excluding the US Supreme Court,[58] the main effort has perforce been before state courts as the jurisdiction primarily seized of tort issues. The *amicus* programme has been particularly prolific in California, increasing in step with the intensity of judicial activism. In 1963-5 CTLA filed briefs in three Supreme Court hearings, in 1973-6 in thirteen Supreme Court and nine Court of Appeal cases, in 1983-5 in eleven and four respectively.[59]

[55] A plea for greater participation in the process of decision-making by English courts, by expanding *locus standi* to make representations to courts, is also made by Bell, *Policy Arguments in Judicial Decisions* 264-6 (1983). For a rare *amicus* brief in English practice (requested from the Official Solicitor by the trial judge) see *Rondel* v. *Worsley*, [1967] 1 QB 443, 445.

[56] See Glick, *Supreme Courts in State Politics*, ch. 6 (1971).

[57] e.g. in actions challenging maximum fee legislation, as in *American Trial Lawyers Association* v. *New Jersey Supreme Court*, 66 NJ 258, 330 A. 2d 350 (1974); 409 US 467 (1973).

[58] e.g. *Norfolk & Western Ry.* v. *Liepelt*, 444 US 490 (1980). Motions to file by ATLA (*eo nomine*) are, *inter alia*, in WESTLAW's data base. Occasionally they are by an affiliate, as in *Liepelt* (Ala. TLA and ATLA). Intervention in lower federal courts is mostly by an affiliate, e.g. *Owen* v. *Diamond M. Drilling Co.*, 487 F. 2d 74 (La. TLA).

[59] The briefs are filed in the names of representative ATLA members without naming the organization itself. The statistics were obtained through

These include all the landmark decisions on products liability, mental disturbance, comparative negligence, and the constitutionality of guest statutes and medical liability reform. Some indication of the attention these briefs receive is found in their arguments being occasionally directly addressed in the judgments.[60] More covert but of no less importance are letters by CTLA representatives in support of petitions for hearing (leave to appeal) to the Supreme Court. The CTLA programme is organized by an *Amicus Curiae* Committee and regularly reported in the *CTLA Forum*. Briefs are written by members, sometimes on commission, often by attorneys working up an appellate practice.

2. THE DEFENCE BAR

The defence bar differs from the plaintiffs' as chalk does from cheese: instead of representing the underdog, it stands up for the corporate establishment; instead of the aggressive, strident, and self-advertising posturing of the trial attorneys, its members mostly cultivate a dignified image of restraint and self-possession, of upholding the existing social and legal system against subversion, untainted by the mercenary and meretricious pursuits of their opponents. Their only common ground is a shared belief in the adversary system.

The different forms of remuneration for the plaintiffs' bar and the defence bar account for the much greater detachment by the latter from their clients' causes. Defence attorneys are not 'principal players'. Significantly, while defence criticism typically focuses more on the conduct of the plaintiffs' attorneys

LEXIS, checked against the law reports. A general sampling revealed that prior to 1927 *amicus* briefs were filed in less than 10% (usually 5%) of Supreme Court cases; hence until the 1970s between 10% and 20%, soaring to 47% in 1978, the last year surveyed. Fernandez, 'Custom and The Common Law: Judicial Restraint and Lawmaking by Courts', 11 *Southw. U. L. Rev.* 1237, 1277 (1979).

[60] e.g. *Barker* v. *Lull Engineering Co., Inc.* 20 Cal. 3d 413, 427, 433; 573 P. 2d 443, 452, 456 (1978); *American Bank & Trust Co.* v. *Community Hospital*, 36 Cal. 3d 359, 382; 683 P. 2d 670, 685 (1984); *Murphy* v. *E. R. Squibb & Sons, Inc.*, 40 Cal. 3d 672, 710 P. 2d 247 (1985); *Vincent* v. *Pabst Brewing Co.*, 47 Wis. 2d 120, 127; 177 NW 2d 513, 515 (1970).

than on that of their clients, it is the conduct of defence attorneys' clients (insurance companies, manufacturers, etc.) rather than that of the lawyers themselves that is usually decried by the other side. Of course, defence lawyers advise their clients on tactics, but in contrast to plaintiffs, defendants are 'repeat players'[61] who mostly call the tune. Thus the opprobrium of 'bad faith' dealings with a plaintiff or his lawyer[62] or of the newly emerging tort of 'malicious defence'[63] is cast on the defendant himself rather than on his attorney.

For the same reason, the public relations contest on the defence side has been conducted principally by the insurance industry, manufacturers, and other target defendants, rather than by their lawyers. Although defence lawyers have formed associations, their activities are principally concerned with refining their skills rather than with cultivating public influence as a branch of the legal profession comparable to ATLA, the plaintiffs' organization. Whie the latter prides itself in being the voice of the 'huddled masses', defendants can—and do—speak up for themselves.

2.1 Ethics

Although the plaintiffs' bar has been the primary target of ethical condemnation, defence attorneys do not altogether escape self-serving temptations. While plaintiffs' lawyers, working for a contingent fee, may tend to invest less effort than the amount needed to secure the maximum benefit for their clients,[64] defence lawyers are tempted to invest more than necessary by padding their time-sheets.[65] Ambulance chasing is another accusation levelled against many plaintiffs' lawyers, but insurance adjusters used to be castigated for beating

---

[61] As distinct from 'one-shotters', as Galanter calls them: 'Why the Haves Come Out Ahead: Speculations on the Limits of Legal Change', 9 *L. & Soc. Rev.* 95 (1974).

[62] *Infra*, at n. 134.

[63] A counterpart to malicious prosecution. See Van Patten and Willard, 35 *Hast. L. J.* 891 (1984).

[64] *Infra*, ch. 6, at n. 65.

[65] See Johnson, 'Lawyers' Choice: A Theoretical Appraisal of Litigation Investment Decisions', 15 *L. & Soc. Rev.* 567 (1981).

plaintiffs' lawyers to the hospital and importuning accident victims into signing quickie releases. Even if some of the more egregious abuses of that sort are now less prevalent, defence interests increasingly cultivate so called 'claims control' strategies like contacting potential claimants or their family lawyers early, before they have committed themselves to a contingent fee contract with a personal injury specialist. Not infrequently, this leads to wrangles between lawyers on both sides of an accident victim, accusing each other in public, or even before ethics committees of the bar, of unethical solicitation of clients.[66]

2.2 Organization

Defence lawyers are much more likely to belong to a largish firm than members of the plaintiffs' bar. But while personal injury claims are not infrequently handled by lawyers in general practice, particularly in smaller communities, the general tendency to concentration has made greater strides on the defence side. Thus many larger law firms specialize exclusively in defence work on behalf of insurance companies, which in turn typically employ one or more law firms for their litigation. A study in Philadelphia indicated than eleven firms handled 40 per cent of the entire trial list for the defence.[67]

The defence bar has parallel organizations to the plaintiffs', although none to equal ATLA in its high public profile or concentration in a single body (with local affiliates). Two organizations deserve particular mention. The DRI–Defence Research and Trial Lawyers Association, founded in 1960, claims 13,000 members among lawyers, claimsmen, adjusters, insurance companies, and target defendants in civil litigation, such as doctors, pharmacists, engineers, manufacturers, and other professionals. Its programme is principally educational, devoted 'to increas[ing] the knowledge and improv[ing] the skills of defence lawyers and to improv[ing] the adversary system of justice'.[68] It conducts seminars and maintains

[66] See Tarr, 'Lawyers Collide Over Post-Air-Crash Conduct', *Nat. L. J.*, 4 Nov. 1985.

[67] Mayer, *supra*, at 258.

[68] *Encyclopedia of Associations* 438 (20th edn. 1986).

research facilities, files of briefs, and an expert witness index. In addition to monographs and occasional position papers, like *Responsible Reform*,[69] it publishes a monthly journal *For the Defense.*

The International Association of Insurance Counsel (IAIC) is an organization of 2,200 members who are practising attorneys engaged in litigation on behalf of insurance companies. It sponsors research projects, an annual week-long trial academy for young trial lawyers, and legal writing competitions. Its most important publication since 1934 is the quarterly *Insurance Counsel Journal*, which, besides useful articles written almost exclusively by practising (usually young) lawyers rather than academics, contains reports from its various sections.

2.3 Political activities

Defence interests have had powerful representation before legislatures no less than before the courts. However, unlike the plaintiffs' bar which has many members in the ranks of legislators themselves, defence concerns are nowadays more generally transmitted by lobbyists to legislators usually sympathetic to conservative causes. Their role, as long as the courts were themselves conservative, was principally to oppose liberal reforms designed to expand liability, such as bills to modify the defence of contributory negligence and to abolish defences like sovereign immunity. On behalf of specific interest groups, occasionally they even sponsored protective legislation, like retraction statutes for libel on behalf of newspapers,[70] Good Samaritan statutes on behalf of the medical profession,[71] and guest statutes on behalf of the insurance industry.[72] But

---

[69] DRI, *Responsible Reform: A Program to Improve the Liability Reparations System* (1969); *Responsible Reform: An Update* (1972).

[70] See Prosser and Keeton, *The Law of Torts* 845 (5th edn. 1984).

[71] Good Samaritan statutes, which lower the standard of legal liability for emergency treatment, were passed at the behest of the AMA in the late 1950s and early 1960s. See Mapel and Weigel, 'Good Samaritan Laws: Who Needs Them?', 21 *S. Tex. L. J.* 327 (1981). Another important success of the AMA was its co-sponsorship of the MICRA legislation to solve the medical insurance 'crisis' (*supra*, n. 40).

[72] See *supra*, ch. 3, at n. 12.

in more recent years, the tactical balance has changed dramatically as liberal judges proceeded to implement the programme of the plaintiffs' bar. Accordingly, defence interests have been looking to the legislature for a more sympathetic ear for their causes. This changed their legislative posture from defence to attack: to reverse court decisions or otherwise to sponsor legislation that would reduce their exposure to liability. Such have been the efforts, already noted, to tame punitive damages and 'cap' damages for non-pecuniary injury.

Defence interests, being more diverse, tend to be represented politically by shifting alliances, depending on the issue involved. Only liability insurers have a consistent stake in opposition to claimants. Much to its chagrin, however, the insurance industry does not enjoy a high measure of public confidence and support, being widely regarded as no less self-serving than its antagonists, the plaintiffs' bar. Its self-interest in maximizing profits is often seen to be at the expense of the consumer. Reform proposals under its banner, however meritorious, are therefore apt to be viewed with suspicion and forfeit their chance to a fair hearing. For public relations purposes, therefore, the insurance industry and its allies like to parade behind more sympathetic masks, like the 'California Citizens' Committee on Tort Reform'[73] or the national 'American Tort Reform Association' (ATRA).[74]

The stake of defence attorneys overlaps but is not always coincident with that of their clients. One cause which notably divides the attorneys from a substantial segment of the insurance industry concerns automobile no-fault.

2.4 Automobile no-fault

As already mentioned, the movement for reforming tort law in the 1960s could not and did not ignore the problem of traffic accidents, which, in the United States as elsewhere, produces the single largest category of casualties[75] and more

---

[73] Its agenda was presented in 1977 under the title *Righting the Liability Balance*.   [74] *Nat. L. J.*, 3 Feb. 1986.

[75] In 1985 1.7m. suffered disabling injuries in motor accidents, 45,000 fatal. The National Safety Council, *Accident Facts* (1985) estimated total cost of auto accidents at $48.6bn., about half the total of all accidents. Of 92,488 deaths by accident, 44,452 were motor-vehicle related.

than half of all tort litigation.[76] But whereas most reform proposals had been addressed to, and many carried out by, liberal courts, even the most activist declined to undertake a fundamental restructuring of the basic rules of motor accidents,[77] in noticeable contrast to the contemporaneous judicial innovations of strict products liability and of comparative negligence (apportionment). In this instance, therefore, the scene shifted from courtroom to legislature, accompanied by a dramatic reversal of the usual roles: the proponents were important segments of the insurance industry, the opponents were the plaintiffs' bar. Even more unusual was that the defence bar for once made common cause with the plaintiffs', in defence of the common law adversary system and against its own traditional clients of the insurance industry.[78]

Proposals for introducing a system of no-fault liability for traffic accidents in America go back to the early decades of this century,[79] stimulated by the model of workers' compensation and early legislation in Europe.[80] But while nothing

[76] In California 55,474 out of 97,000 tort filings in 1983–4 related to motor vehicle accidents (Judicial Council of California, *1985 Annual Report* 106), but there has been a substantial decline in 'contested dispositions' from 1,500 in 1974–5 to 909 in 1982–3 to 728 in 1983 (ibid., at 120). The decline is also attested by the Rand jury study which recorded an average of 48% of total trials between 1960 and 1980, but a steady decline to a mere 38% in the late 1970s (Shanley and Peterson, *supra*, at 66–9). In Chicago, automobile cases amounted to 66% of all jury trials, with a negligible decline over the same period (ibid.; Peterson and Priest, *The Civil Jury: Trends in Trials and Verdicts, Cook County, Illinois, 1960–1979* 7 (1982).

[77] *Supra*, ch. 2, at n. 96.

[78] DRI's *Responsible Reform* 32 opposed no-fault in 1969. By 1972 (*Responsible Reform: An Update* 11–12) it proposed a limitation of general damages to a sum proportionate to necessary medical expenses except in cases of death, dismemberment, permanent disability, disfigurement, or loss of function. That list corresponds to thresholds under no-fault plans.

[79] The most systematic was the so-called Columbia plan of 1932, presented by a Committee to Study Compensation for Automobile Accidents to the Columbia University Council for Research.

[80] Germany introduced a system of strict liability in 1909. It excludes non-economic damages (which can be claimed in tort on proof of negligence) and is subject to a current limit of 500,000DM (approximately US $250,000). See generally Tunc, *Intern. Encycl. Comp. L.*, vol. xi, ch. 14 ('Traffic Accident Compensation: Law and Proposals').

came of these radical suggestions, significant changes gradually modified the traditional fault system in the direction of more widespread and effective compensation of victims. Entitlement became broadened as a result of the abolition of immunities, especially the interspousal and interfamily immunities, followed by the repeal of guest statutes, both of which successively qualified passengers as claimants.[81] Added to these reforms was the replacement of the defence of contributory negligence by comparative negligence (apportionment) which officially entitled negligent plaintiffs to some (reduced) compensation.[82]

Not only liability law, but also insurance law and practice improved the lot of plaintiffs. A number of jurisdictions enacted compulsory insurance, though far too few of them and in amounts which eroding inflation has rendered absurdly inadequate.[83] The remainder enacted 'financial responsibility laws', cancelling the licence of culpable drivers who are uninsured or cannot furnish security—on the model of dangerous dogs to whom the law affords 'one free bite'. In the absence of official funds for victims of uninsured or unidentified drivers, insurers offer uninsured drivers coverage[84] which allows an insured victim to claim from his own insurer in case his culpable injurer is uninsured—an interesting free-market solution for dealing with the significant percentage (more than 10 per cent) of uninsured drivers, among them a large proportion of the most risk prone who cannot afford the high premiums on account of their dismal accident records.[85]

---

[81] *Supra*, ch. 3, at n. 12.

[82] *Supra*, ch. 2, at n. 56. Six states retain the traditional common law defence (Ala., Miss., Tenn., Ariz., Ind., NC.); none has no-fault and the first 3 even lack compulsory liability insurance.

[83] Until the wave of no-fault legislation, only New York and North Carolina demanded compulsory third-party insurance. Accompanying no-fault plans, 11 states made third-party insurance compulsory and so have a few traditional states like California. The required coverage ranges from $10,000 to $25,000. See the list on p. 77 of US Department of Transportation, *Compensating Auto Accident Victims* (1985), hereinafter abbreviated DOT.

[84] Forty-nine states require that it be offered, 19 that motorists purchase it: *Automobile Accident Compensation*, vol. iv: *State Rules* 16 (1985). Some states, like California, now require also under-insured motorist coverage. New York has a central fund, the Motor Vehicle Accident Indemnity Corp.

[85] Rates range from near zero to 30% (California: 17%): *Automobile Accident Compensation*, supra, at 18–19.

Lastly, insurance cover can no longer as a rule be denied to victims because of policy-infirmities, such as misrepresentation, or because of the insured's supervening bankruptcy.[86]

Thus, both the law and insurance practice have done a good deal, especially in the more recent past, to improve the compensation record. Awards in motor accident litigation have remained relatively modest in comparison with the high-flying medical malpractice or products liability awards, probably because juries have been made well aware of the long-term effect of inflated verdicts on the level of their own insurance premiums.[87] Nevertheless, a large proportion of claimants receive some compensation, although its proportion to their total economic loss declines precipitately with increasing gravity of the injury. Thus while the recovery for minor injuries tends to exceed the claimants' economic loss, that for catastrophic injuries falls disastrously short.[88]

American no-fault automobile plans developed organically out of the progressive transformation of accident insurance, first- and third-party, from private to a public function. Their model was not the European one of strict liability backed by compulsory insurance, but that of voluntary first-party insurance available to motor-car owners for injury to them-

---

[86] This has been accomplished by a combination of legislation and changing the formulation of Bureau policies.

[87] The Chicago study by Peterson and Priest, *supra*, at 54–7, reported that between 1975 and 1979 38% of plaintiffs' awards were less than $3,000, compared with 22% between 1960 and 1964. The median has also steadily decreased. Half of them were less than $5,000. Two explanations are offered: jurors might be less generous or different kind of cases are brought to trial. The San Francisco experience was closely parallel: Shanley and Peterson, *supra*, at 14–16. Both Illinois and California are traditional tort jurisdictions without no-fault.

[88] The Department of Transportation in 1970 estimated that 47.7% of seriously or fatally injured victims recovered at least in part (*Economic Consequences of Automobile Accident Injuries* 47 (1970)); a 1977 AIRAC study based on people with auto insurance suggested 64%. For the overpayment of slight injuries see Conard *et al.*, *Automobile Accident Costs and Payments*, ch. 5 (1964). $24.4bn. was paid by insurers to auto victims in 1985, average personal injury claim being $6,166 (Kakalik and Pace, *supra*, App. D 137–40).

The English record is apparently worse: 29% according to Harris *et al.*, *Compensation and Support for Illness and Injury* 51 (1984); *Pearson Report*, vol. ii, para. 201 (25%).

selves and their passengers.[89] Instead of remaining voluntary,
the insurance simply became mandatory. This simple structure
had many advantages. First, it naturally encompassed coverage
not only for third parties but also for the driver-owner, in
contrast to the clumsy device necessitated by the liability
model of the recent French reform.[90] Second, it did not require
the setting up of a new bureaucracy to administer a central
fund but could continue to function through private insurance.
This feature made it attractive to segments of the insurance
market and thereby moved into the realm of political feasibility.
Third, the first-party model severed all links, symbolic no less
than substantive, from tort liability. This removed, or at least
palliated, the adversary element in the relation between
claimant and insurer. Other possibilities flow from the fact
that, whereas the premiums of conventional liability insurance
are based on the potential risk to others, those of first-party
insurance are based on risk to oneself. This makes the latter
more cost-efficient in that the beneficiary can be left some
choice in the amount of coverage needed. For example, if he
already maintains medical insurance for himself and his
family, or other protection against disability, he can opt for
appropriate deductibles and thereby lower the cost of insurance. In other words, the insurance can be tailored to
individual, predictable needs, in contrast to liability insurance
which has to cover the risk of injury to high as well as low
earners and therefore incidentally works against the poorer
section of premium-payers. On the other hand, it had to
confront the problem of the collision between truck and small
car. Small cars, being more vulnerable, would have to bear
the high cost of collision with trucks, but this inequity has
been counteracted by permitting subrogation against the latter
or by imposing a surcharge on them.[91]

[89] It is somewhat paradoxical that while Europeans (including the British)
after World War II replaced the third-party tort model for workers' injuries
by incorporation into social insurance but retained it for automobile plans,
the American experience has been just the reverse: retention of the original
British model of workers' compensation, but first-party insurance for automobile plans.

[90] Law of 5 July 1985. See Brousseau, *La Loi Badinter* (1985).

[91] Five states have created a right of subrogation based on fault, 7 states
regardless of fault. See DOT *supra*, at 118–19.

Keeton and O'Connell's *Basic Protection for the Traffic Victim: A Blue Print for Reforming Automobile Insurance* (1965) furnished the inspiration and the model for the no-fault insurance legislation which came to be adopted in the ensuing years in about half the American jurisdictions. Although differing markedly from each other, they can be classified according to certain predominant characteristics. The principal division is between those that do and those that do not restrict tort claims. Sixteen states belong to the first ('no-tort'), eight to the second ('add-on') category, while twenty-eight retain the traditional system of tort liability coupled with third-party insurance. Since tort liability is not completely abolished in any state, as it is in Quebec[92] and Israel,[93] liability insurance continues to remain alive everywhere.

The effectiveness and success of the various no-fault plans are conditioned by the interplay between two variables: the level of no-fault benefits, and the 'threshold' which a victim must be able to cross before qualifying for a tort claim.[94]

First, no-fault benefits vary from $1,000 in South Carolina and $2,000 in Massachusetts and Utah to unlimited medical expenses in Michigan and New Jersey or $100,000 in the District of Columbia and $50,000 in New York.

Secondly, three basic types of thresholds are variously employed. All but three of the no-tort statutes employ thresholds expressed in dollars of medical expense, ranging from $500 to $5,000. A second type of threshold is linked to days of disability. Most successful have been the 'verbal' thresholds, such as 'permanent' and/or 'serious' disability, fractures, disfigurement, or death. All but one statute contain verbal thresholds, three exclusively. Experience has shown a clear relationship between the percentage of cases removed from the tort system by the threshold and the total amount of money paid to all victims (including tort claimants), without increasing insurance premiums adjusted for inflation.[95] The

---

[92] See Baudouin, 31 *Rév. Intern. Dr. Comp.* 381 (1979).

[93] See Yadin, 28 *Rév. Intern. Dr. Comp.* 475 (1976).

[94] The following account relies principally on DOT, *supra*, and on the Rand studies *Automobile Accident Compensation*, vols. i–iv (1985). Both sources are based on the same data.

[95] DOT, *supra*, at 95–107. According to a Rand study (Rolph, *Automobile Accident Compensation*, vol. i, 19 (1985)), a threshold of $200 in medical costs

three states with an exclusively verbal threshold recorded lower insurance costs than would have been without no-fault. Conversely, very high no-fault benefits accompanied by only very modest limitations on tort claims resulted in large premium increases, leading Pennsylvania to repeal no-fault altogether in 1984. By contrast, the combination of either substantial benefits and thresholds[96] or low benefits and thresholds[97] resulted in 'balance'.

That verbal thresholds produce more effective fiscal restraints is demonstrated by the experience in New York and Florida which dramatically attained balance after relinquishing dollar thresholds in the late 1970s. The weakness of dollar limits lies in their erosion by inflation and the temptation to pad medical costs. Too few states have high dollar thresholds to permit any confident conclusion whether dollar amounts could be an effective alternative to exclusive verbal thresholds.[98] Calculations based on the experience in high benefit states (Michigan, New Jersey) justify the prediction that any total elimination of tort claims would result in very large savings, in turn permitting insurance premiums to be 'held stable or even lowered even if benefits were increased to high levels and even if pain and suffering benefits were made available on a no-fault basis'.[99]

How effective have no-fault plans been, in comparison with traditional tort law, in delivering compensation for medical and rehabilitation expenses? How many more claimants have received some compensation and how have the catastrophically injured fared? Two studies have researched the first question.[100] One shows that the claim frequency in 1983 for no-fault was 1.84 per hundred cars insured in each of the no-tort states and 1.37 in add-on states, compared with 0.92 in traditional jurisdictions. Thus almost twice as many victims received some

(net of hospital, NJ) eliminated 48% of tort claims, $500 (incl. hospital, Mass.) 60%, $750 (Pa.) 75%, and Michigan's verbal only threshold, 89%.

[96] e.g. New York ($50,000), Minnesota ($30,000).

[97] e.g. Massachusetts ($2,000).

[98] The Minnesota experience ($4,000 since 1978) is mixed, but early DC results for its $5,000 limit introduced in 1983 are encouraging.

[99] DOT, *supra*, at 105.

[100] Ibid., at 73–8.

compensation under no-fault. Another study in 1977 calculated that 65 per cent of all automobile victims would have received some compensation if no-fault had been available, compared with 47.7 per cent whom an official 1970 study found receiving some compensation under the fault system.

According to a 1982 estimate the average medical cost for catastrophic injuries (above $100,000) was $408,700. Only Michigan, New Jersey, and Pennsylvania's unlimited medical benefits would have been adequate; in most others there would have been a shortfall of at least $350,000. But so also there would have been in the traditional states if regard is had to the minimal limits of required third-party liability insurance, which nowhere exceeds $25,000 and is mostly between $10,000 and $15,000. In individual cases the shortfall may be less, because the motorist may voluntarily have had greater insurance or other assets. But the overall performance of the fault system is even worse because a large proportion of the pay-out is for non-economic injury, thereby reducing further the amount attributable to medical care and rehabilitation.

As indicated in the preface to this section, no-fault legislation has been the occasion for a rare consensus between both sides of the trial bar in defence of the acclaimed heritage of the adversary system. While the defence lawyers have on the whole been content to let the plaintiffs' bar carry the torch in opposing reform bills, both have a financial stake in the maintenance of the traditional tort system. The plaintiffs' bar is less opposed to add-on statutes which mop up minor claims but also tide over more severely injured plaintiffs until their tort claims are satisfied.[101] Their hostility increases with higher benefits and more rigorous thresholds: in other words, with the greater adequacy of the scheme to meet the compensation needs of casualties. The fact that in more than half the states no-fault bills were never enacted (none since the mid-1970s) and that in most others benefits and thresholds are low reflects the political influence of the plaintiffs' bar in many legislatures.

---

[101] A recent study concluded that only 10% of no-fault claimants employ lawyers to assist with their no-fault claims (others retain lawyers for a con-current tort claim), whereas half of tort claimants employ lawyers: Rolph, *supra*, at 24.

Their propaganda focuses on rising insurance premiums without making allowance for inflation or adverting to the success of no-fault plans 'in balance'.

While the defence bar has remained a staunch defender of the fault system,[102] the insurance industry, their traditional client, has furnished the chief promoters of no-fault bills. Inflation was the industry's chief ally, as medical and motor-car repair costs in the period 1965-75 were rising at a much faster rate than the average cost of living and therefore threatening to price insurance out of the market and reduce profits. Some insurance companies remained loyal to their belief in the advantages of the adversary system, but others (among them the most powerful) turned away, reflecting respectively individualized and aggregative approaches to risk selection.[103] The United States Department of Transportation has consistently supported the no-fault effort.[104]

3. INSURANCE ADJUSTERS

All but a tiny fraction, less than 5 per cent, of all successful tort claims terminate in a negotiated settlement rather than a judicial adjudication.[105]

Public policy strongly supports settlement in the interest of clearing dockets, especially for routine cases such as most tort

[102] *Supra*, n. 78.

[103] Spearheading reform was the American Insurance Association's *Complete Personal Protection Plan* (1968). The AIA is a trade group which includes a large number of prestigious stock companies. See Ross, *Settled Out of Court* 274-7 (1970).

[104] The Department of Transportation, at the request of Congress, published a number of studies on the automobile accident problem and recommended guidelines for no-fault plans. See *Motor Vehicle Crash Losses and Their Compensation in the U.S.* (1971). Its latest follow-up study is DOT, *supra*.

[105] Ross found in a 'closed claims' file of one major casualty insurer that of 2,216 personal injury claims all but 93 were settled without trial: Ross, *supra*, at 179. The literature generally assumes that at most 5% go to trial. According to Judicial Council of California, *1985 Annual Report* 118, 95% of all Superior Court dispositions in 1983-4 were pre-trial, 3% by uncontested trial, and only 2% by actual trial. Compared with 1974-5 the increase of 'dispositions' for motor vehicle cases was only 17%, for all personal injury cases 28% (ibid., at 116).

claims. Ironically, several highly negative features of the American adjudicative process contribute to that desirable goal. Chief among them are the calamitous delay in getting to trial, the procedural complexity and chicanery open to litigants, and the uncertainty of legal outcome implicit in the jury system. English courts, as already noted, are adamantly opposed to jury trial on damages in the belief that judicial tariffs facilitate settlements.[106] Others, however, believe that the uncertain dollar value of 'general damages' tends to induce plaintiffs to settle because they are risk-averse.[107] Besides these unintended incentives due to the inadequacies of the adjudicative system, settlement is deliberately encouraged by pre-trial discovery, which eliminates surprise, and the pre-trial conference at which judges cajole litigants into settlement.[108] Both, however, are liable to abuse. As a result, access to trial is severely impeded. This explains why the number of completed, even of commenced, trials has barely increased over the years despite the spectacular rise in court filings.[109] The tort explosion, if it be such, in our generation has been an explosion of claims rather than of trials. It is open to question whether such rationing serves the public interest, especially as it tends to distort the settlement process as the following discussion illustrates.[110]

[106] *Supra*, ch. 4, at n. 100.

[107] But Ross, *supra*, at 163 counts this as a major cause of failure to reach a settlement. For the wide variation in the proportion between special and general damages see Rolph, *supra*, at 20–2. Although adjusters are widely credited with using rules of thumb ('multiples'), systematic investigation does not bear out any regularity.

[108] It has also been proposed, as in England, to make a litigant liable for his opponent's attorney fees if he receives a judgment less favourable than a settlement offer he rejected. 98 FRD 339, 361–3 (1983). Current Rule 68 of the Federal Rules of Civil Procedure merely mulcts such a litigant with the extra legal costs, excluding attorney's fees.

[109] From the early 1960s to the mid-1970s San Francisco averaged 290 jury trials p.a.; thereafter the number declined to a low of 158 in 1978, rising to 210 in 1979. Chicago experienced a very slight increase over the same period. Shanley and Peterson, *supra*, at 19–20.

[110] See Alschuler, 'Mediation With a Mugger: The Shortage of Adjudicative Services and the Need for a Two-Tier Trial System in Civil Cases', 99 *Harv. L. Rev.* 1808 (1986). See also Fiss, 'Against Settlement', 93 *Yale L. J.* 1093 (1984), who equates settlement in civil cases with plea bargaining in criminal cases and contends that its effect is most coercive on disadvantaged

The principal actors in the process of negotiating settlements on the defence side are insurance adjusters. Defence lawyers do not usually become involved unless and until the insurance staffers have failed to bring about a settlement on their own. Herein then lies an important difference in the respective participation of lawyers on either side. A large proportion of successful claimants are represented by lawyers from a fairly early stage; indeed unless they are, many of the injured give up sooner or later.[111] Moreover, there is evidence that, at least in cases of graver injuries, the amounts received, especially non-economic damages, are considerably enhanced by legal representation.[112] Most effort by plaintiffs' lawyers is therefore expended in settlement negotiation rather than in the courtroom, notwithstanding their self-flattering description as 'trial lawyers'. Their opponents, however, are primarily insurance adjusters, not members of the defence bar.

The process of negotiation has a *Gestalt* very different from that of adjudication. The object being to reach agreement with the opponent, advocacy of one's cause must be accompanied by willingness to compromise. While the negotiations are carried out 'in the shadow of the law',[113] inevitably the outcome will be influenced by other forces, such as the relative balance of bargaining power, differing attitudes towards risk, and the litigants. The stress of uncertainty on the lawyer–client relationship is discussed by Rosenthal, *supra*, ch. 3.

[111] e.g. the New York study by Franklin, Chanin, and Mark, *supra*, at 13, 42, reported that the frequency rate of recovery by represented insurance claimants was 90%, that of the unrepresented 65%.

[112] In Ross's sample (*supra*, at 193) 1,601 claimants were unrepresented. Representation significantly affected the amount of recovery, suggesting a rate from 5 to 20 times as high. This outcome unquestionably reflects selection in that attorneys act as gatekeepers in their choice of clients. Thus representation has been shown to be much more likely in cases with at least moderate injuries: ibid. 193–8. See also *infra*, ch. 6, n. 166. Even allowing for the attorney's fee, it is generally worth a client's while to be represented: ibid.: Rolph, *supra*, at 24–6.

[113] A term popularized, if not coined, by Mnookin and Kornhauser, 'Bargaining in the Shadow of the Law: The Case of Divorce', 88 *Yale L. J.* 950 (1979). While negotiation is a less norm-based process than adjudication, it is neither norm-free nor dominated exclusively by horse-trading: A third alternative, allowing recovery of attorney fees only to successful plaintiffs, is analysed and recommended by Leubsdorf, 'Recovering Attorney Fees as Damages', 38 *Rutgers L. Rev.* 439 (1986).

uncertainties and cost of actual trial. Laurence Ross, in his classical study, *Settled Out of Court*,[114] has given a detailed description and analysis both of the process and of its players. A few salient features which bear directly on the major theme of the present work deserve special emphasis.

Admittedly, insurance adjusters are basically a conservative lot, most of whom relish the adversary role *vis-à-vis* plaintiffs' attorneys and approach each claim in the hope of denying it or making as low a payment as possible. If faced with an unrepresented claimant, their aim will be to dissuade him from engaging a lawyer but usually to offer payment for out-of-pocket expenses.[115] Once a lawyer appears on behalf of a claimant, the stakes change dramatically, and so do the rules of the game.[116]

But the claimant's legal representation is not the only modifying influence. There is, first, the organizational pressure, emanating from the adjuster's hierarchy of supervisors, to close cases promptly. In relatively small claims the cost of protracted negotiation is out of all proportion to any possible savings and counsels quick payment, with the result that determined demands are usually met without going too deeply into the merits. An adjuster's success rate is measured by the number of files closed.

Next are the 'danger' and 'nuisance' factors. Although liability may be far from clear, the size of damages may counsel compromise. If the damages suffered are grave, juries, as already noted,[117] are prone to feel proportionately greater compassion for the victim. This in turn increases the 'danger value' of the claim.[118] Perhaps less openly admitted by adjusters is the 'nuisance value' of routine claims, which is explained by the pressure previously mentioned to close files expeditiously. It is most prominent in the adjustment of collision and other property damage claims which, especially

---

[114] *Supra*, n. 103. For an economic analysis see Note, 'An Analysis of Settlement', 22 *Stan. L. Rev.* 67 (1969). A parallel English view is by Phillips and Hawkins, 'Some Economic Aspects of the Settlement Process: A Study of Personal Injury Claims', 39 *Mod. L. Rev.* 497 (1976).

[115] Ibid., at 166–72.

[116] Ibid., at 116–33.

[117] *Supra*, ch. 4, at n. 41.

[118] See Ross *supra*, at 199–204.

in small amounts, are usually paid off without even verifying the damage.[119]

On the other hand, the dynamics of negotiation also contain negative consequences for claimants. For settlements are also discounted by the claimants' own uncertainties of winning at trial (and how much), with the result that their recoveries in all but the clearest cases will be less, often substantially less, than a favourable jury verdict might offer in an unpredictable future. We noted earlier the tendency of smaller claims to be compensated in excess of the economic loss, while larger claims suffered increasing shortfalls.[120] The reason for the first is that the effect of collateral benefits and awards of non-economic damages are proportionally greater at low levels. The reason for the second, however, is a failure not of the negotiating process, but of the low limits of the typical motor-car defendant's insurance cover. For reasons to be explained below, insurance adjusters are under heavy pressure to accept settlement offers within the insurance limits; beyond those limits plaintiffs will usually go empty-handed unless they can tap some other deeper pocket.

Other discount factors are the delay in awaiting trial and the risk-averseness of the typical claimant. Delay in the judicial disposition of civil claims is one of the notorious American maladies.[121] In Los Angeles the median time between filing and trial of a civil action lengthened from four and a half months in 1942 to forty-one and a half months in 1982.[122] Even in small cities a waiting period of thirty months has become routine. Such delay works almost always in favour of the defence. Moreover, the coercive effect of delay in encouraging victims to settle early for less than their claim increases in proportion to the size of the claim, much to the disadvantage of the most seriously injured. This furnishes

[119] Ibid., at 204–11.

[120] *Supra*, n. 88.

[121] See Rosenberg and Sovern, 'Delay and Dynamics of Injury Litigation', 59 *Colum. L. Rev.* 1115 (1959); Zeisel, Kalven, and Buchholz, *Delay in Court* (1959).

[122] Selvin and Ebener, *Managing the Unmanageable: A History of Civil Delay in the Los Angeles Superior Court* 17 (1984). Rosenthal, *supra*, at 89 reported a delay in Manhattan of 34 months between a case being put on the jury calendar and tried.

another explanation why large losses are relatively under-compensated. Moreover, the record shows a strong negative correlation between delay and the amount recovered (and, presumably, claimed).[123] Insurance companies have been accused of systematically exploiting the vulnerability of the more seriously injured claimants by artificial delays, from which the companies derive the added advantage of earning income on their invested funds in the meantime and of seeing claimants become increasingly less able to sustain a successful trial with the flux of time. A partial corrective would be to repeal the common-law rule against pre-judgment interest on unliquidated damages, a reform adopted in England and some Commonwealth jurisdictions,[124] but not in many American states.[125] Clearly, the rationing of adjudicative services by queue has a distinctly anti-plaintiff bias.

Finally, not to be minimized is the different attitude towards risk by claimant and insurer. Victims are generally risk-averse, insurers are risk-neutral. Settlement offers a known award in place of an unknown probability of winning at trial. Besides involving extra costs, litigation involves a gamble that may be totally lost.[126] For the plaintiff, therefore, everything is at stake with only one throw of the dice. Indeed, the pressure is magnified by its effect also on his attorney, who is especially averse to litigation since it is he who will bear the cost of the trial, whatever its outcome, besides his own fee being contingent on recovery.[127] By contrast, the insurance company can contemplate the choice between certainty and the gamble with equanimity. Each case is merely one of many, the cost being spread among all. Moreover, the adjuster has no personal stake in the outcome; indeed, many seem to prefer a judicial

---

[123] Ross, *supra*, at 64-6. Of those recovering $1,000 or less, 68% closed within 1 year and 85% within 2. But of those recovering more than $3,000 only 20% closed within 1 year and 40% within 2. Representation by an attorney significantly lengthens the average waiting period, no doubt at least partly because such cases would be more complex: Rolph, *supra*, at 24-6.

[124] England: see *McGregor on Damages*, ch. 14 (14th edn. 1980); Australia: see Luntz, *Assessment of Damages* § 11.02 (2nd edn. 1983).

[125] See Fleming, 'Tort Damages for Loss of Future Earnings', 34 *Am. J. Comp. L.* (Suppl.) 141, 155 (1986).

[126] Ross, *supra*, at 214.

[127] *Infra*, ch. 6, at n. 66.

adjudication to assuming a stigma within their organization for agreeing to a very large settlement.[128]

Yet, making allowance for all these negative pressures, the remarkable fact remains that most insurance claims receive some payment. According to an investigation of New York automobile cases in 1957, 84 per cent of all victims who made claims against insurance carriers achieved some recovery;[129] according to another covering all third-party liability claims against one particular large carrier, the success rate by plaintiffs was 66 per cent.[130] Evidently, even claimants with weak credentials, many of whom would most probably have lost at trial, thus benefit from the law of torts. It explains why even before the introduction of comparative negligence a very large proportion of persons injured in vehicle collisions succeeded in recovering some damages, more as a result of settlement than of jury manipulation. However, both perform an identical function in differentiating the law in operation from the law in books. The fact, therefore, that the law as practised in the 95 per cent or more of serious automobile claims operates with a powerful bias in favour of compensation should dispel widely held misconceptions not only about the success rate of tort claims but also regarding the performance by insurance companies in their imputed role of compensating victims. Accordingly, the apparent belief by many liberal judges that insurance companies engage in systematic obduracy and chicanery in resisting their obligations is largely based on a misunderstanding of the powerful market forces that control the negotiating process. Their programme for improving insurance practices, about to be described, should be evaluated in the light of these findings.

---

[128] Ross, *supra*, at 221.

[129] Franklin, *supra*, at 13. But the alacrity of insurers to settle varies greatly. Rosenthal, *supra*, concluded that 22.9% had a pattern of settling 'reasonably early', 42.9% 'reasonably late', 12% 'never reasonable'.

[130] Ross, *supra*, at 182. Indicative of the fact that liability rules strongly, but far from conclusively, determine the outcome, was Ross's calculation that of 386 claims judged to entail liability 91% were paid, of the 232 claims judged to entail no liability 34% were paid, but received a much smaller sum; and of 168 claims judged questionable 80% were paid in amounts about equal to the first group: ibid. at 183-4.

3.1 Settling in good faith: 1

The campaign of censure against insurance companies has pursued two targets. Plaintiffs' organizations, in retaliation for legislative efforts to roll back gains secured in liberal courts, have sought to introduce more effective controls of the insurance industry by requiring public disclosure of their finances and regulation of insurance premiums by state insurance commissioners.[131] The so-called insurance crisis, in their view, was not so much the result of an over-expansive law of torts as of mismanagement and greed by the casualty insurance industry. More wounding has been the judicial attack on insurance practices, with a view to enforcing a more compliant attitude towards claimants. The key was found in the insurer's covenant of good faith.

In the United States, as already noted, the mandatory cover for third-party automobile liability insurance is invariably limited, usually to a rather small sum like $15,000/30,000[132] which has not been updated despite sustained inflation and bears no realistic relation to current medical costs and lost wages incurred as the result of substantial injuries. What is more, all too many automobile owners fail to procure higher cover despite their exposure to liability in six and seven figures. What sanctions are there against an insurer turning down a settlement offer by the plaintiff within the insurance limits and thereby exposing his client, the insured, to personal liability on a later verdict in excess of those limits? The likelihood of abuse is of course controlled in the first instance by the strong incentive for insurers to settle whenever the

[131] W. Va. was the first state, in 1986, to require malpractice insurers to give 'due consideration' to each policyholder's past loss experience, to open their books and provide explanation of how they derive their rates. Five national insurance companies were reported to have promptly notified their clients of cancellation. *Wall St. J.*, 15 Apr. 1986, p. 6.

There is no uniformity as regards rate regulation. Many states do not even require filing of rates with the state insurance commissioner. In general, rate regulation used to focus more on rate adequacy than rate equity, but some states now endeavour to hold rates down. See *Automobile Accident Compensation, supra*, at 20–3.

[132] i.e. $15,000 for each claimant, $30,000 for each accident. Colloquially referred to as '15 and 30'. In 1967 when California prescribed this cover, the daily hospital cost was $65, now it is over $1,000.

likely recovery and the cost of trial exceed the policy limits. Beyond that, the conventional answer, still adhered to in many states,[133] has been to make the insurer liable for the excess only on proof of malicious refusal to settle in violation of his implied covenant of good faith.

The California Court fired the first salvo in its campaign to reform insurance practices by ruling that the covenant of good faith called for reasonable care in the interest of the client, thereby replacing malice by negligence as the appropriate standard.[134] Indeed, it was plausible to go one step further and resolve the insurer's inevitable conflict of interest by putting him at absolute risk in his decision whether or not to accept such a settlement offer.[135] What gave the decision its real teeth, however, was related to the measure of damages. First, the plaintiff's compensatory damages were measured by the amount of the verdict rendered against him rather than of his actual, possibly lesser loss, having regard to his limited financial resources which were taken in execution by the successful tort claimant. Thus in the leading case the plaintiff, owner of an apartment house, was sued by a tenant who was hurt on an outside stairway and claimed $400,000 for her physical and psychic injuries. The plaintiff had a $10,000 general liability policy with the defendant insurance company which refused payment for the mental injury claim despite medical evidence on both sides of the causation issue. The tenant's demand for $10,000, later $9,000, was rejected although the insured owner offered to pay $2,500 of it. The jury awarded $100,000, the insurer paid $10,000 and the owner's building was eventually taken in execution to satisfy the tenant's judgment. The owner in turn sued her insurance company and recovered, not the value of her property taken in execution, but the amount awarded to the tenant minus $10,000. While this measure may be vulnerable on logical

---

[133] See Keeton, *Manual of Insurance Law* 510–11 (1971): also annot. 40 ALR 2d 168 (1955).

[134] *Crisci* v. *Security Insurance Co.*, 66 Cal. 2d 425, 426 P. 2d 173 (1967).

[135] Ibid., at 431, 426 P. 2d at 173 (Peters J.). In case of denial of coverage, as distinct from refusal of settlement offers, liability is strict, the insurer taking its chances.

grounds,[136] the manœuvre was obviously intended to benefit the tort claimant who was, of course, not a party to the contract. Indeed, henceforth, it became the practice for the insured to assign his claim against the insurance company to the tort plaintiff, typically in return for a promise to co-operate with the latter in future litigation over liability.[137] Such arrangements are condoned by the courts despite their obvious invitation to collusion at the expense of the insurance company.[138]

Secondly, the court added a further incentive by sanctioning damages for emotional distress, reasoning by reverse logic (from conclusion to premiss) that a 'malicious' breach of contract was a tort and that in any event, the aim of the contract being to protect the promisee against anxiety, such damages were exceptionally appropriate even for breach of contract. Subsequently, this conclusion was reinforced by allowing an additional award of punitive damages on proof of conscious disregard of the client's interests.[139] Since the benefits of this bonanza are in large measure shared by the original tort victim, one wonders whether her attorney would not rather have the insurer refuse her settlement offer and

[136] See Keeton, *supra*, at 514–20. It has been defended on the ground that measuring liability by leviable assets might invite insurers to be less considerate to poor policyholders than to rich ones: Note, 'Liability Insurers and Third-Party Claimants: The Limits of Duty', 48 *U. Chi. L. Rev.* 125, 130–1 (1981).

[137] The claim for 'excess liability' is treated as contractual and therefore assignable even in states like California which do not permit assignment of tort claims for personal injuries. But the claim for emotional injury would not be assignable. See annot. 12 ALR 3d 1158.

[138] A *pessimum exemplum* is *Samson* v. *Transamerica Insurance Co.*, 30 Cal. 3d 220, 636 P. 2d 32, 178 Cal. Rptr. 343 (1981). So-called 'Mary Carter' agreements go one step further by stipulating that the settling tortfeasor's liability decrease in proportion to the award recovered from the other(s): *infra*, n. 142.

[139] e.g. *Neal* v. *Farmers Insurance Exchange*, 21 Cal. 3d 910, 582 P. 2d. 980 (1978) ($1.5m., reduced to half; upheld despite egregious misconduct by plaintiff's attorney). Followed in several other jurisdictions: see 85 ALR 3d 1211 (1978). The bad faith formula has become a Trojan Horse, inviting punitive damages even in the absence of abusive behaviour. To expect juries to keep the two distinct is asking too much. Yet the California Supreme Court, during its radical era, had a record of not correcting plaintiffs' verdicts.

hope for a substantial verdict than accept the offer within the low policy limits.[140]

Eventually, this Byzantine judicial exercise was capped by the California Court conferring a direct tort claim on both the insured *and* the injured claimant, based on a general provision in the Insurance Code, promoted by the National Association of Insurance Commissioners (NAIC), which prohibits a variety of unfair claims practices, including failure to attempt 'in good faith to effectuate prompt, fair and equitable settlements of claims in which liability has become reasonably clear'.[141] The court rejected arguments against implying a third-party cause of action for damages besides the cease-and-desist orders expressly provided for, and against basing a claim on a single violation. Its only concession was that such a claim should be determined after trying first and separately the claim against the insured.

Few other states have so far been beguiled into adopting as extreme coercive measures against insurers at the cost of obscuring the traditional boundaries of contract and tort.[142] These developments may be claimed to have had a profound and beneficial effect on settlement practices, but in turn contributed their share to the escalation in premiums associated with the so-called insurance crisis. In the end, the cost of greater benefits has to be borne by the benefiting public. Less drastic alternatives are available. Perhaps the most promising, to be considered later, is the exchange of punitive damages for reasonable attorney fees against a recalcitrant insurer.[143]

---

[140] It has become routine for plaintiffs' attorneys to send a 'bad faith' letter accompanying an offer to settle within policy limits.

[141] *Royal Globe Ins. Co.* v. *Superior Court*, 23 Cal. 3d 880, 592 P. 2d 329 (1979); *Pray* v. *Foremost Ins.*, 767 F. 2d 1329 (9th Cir. 1985) (liable although verdict far below settlement offer refused); but *Sych* v. *Ins. Co. of N. America*, 173 Cal. App 3d 321, 220 Cal. Rptr. 692 (not liable if plaintiff failed against tortfeasor). Followed in W. Va. and Mont.; Fla. enacted a specific cause of action. On statutory penalties generally, see 33 ALR 4th 579 (1984).

[142] e.g. *A & E Supply Co.* v. *Nationwide Mutual Fire Ins. Co.*, 798 F. 2d 669 (4th Cir. 1986), rejecting not only the third-party claim under the model Insurance Code, but also the bad-faith claim for punitive damages. For detailed criticism of the third-party claim see Note, *supra*, n. 136, at 134–56.

[143] See Curtis, 'Damage Measurements for Bad Faith Breach of Contract: An Economic Analysis', 39 *Stan. L. Rev.* 161 (1986).

3.2 Settling in good faith: 2

In another context the requirement of 'good faith' in settlement has been viewed with less enthusiasm by plaintiffs and their attorneys. Since it reveals a good deal about professional tactics and ethics, it is a story worth telling in conclusion.

The question concerns the effect to be given to an undervalue settlement with one of several co-tortfeasors. Under the rule embodied in the Uniform Contribution Act of 1956, followed in many States, a good faith settlement releases the settling defendant from all further liability, including claims for contribution from co-tortfeasors. Moreover, the plaintiff need only give credit to other defendants for the amount recovered in the settlement, not for the settlor's share of apportioned fault. Thus the risk of an undervalue settlement is not borne by the plaintiff but by the remaining defendants who were not privy to the settlement.

Both aspects of this solution—exempting the settling defendant from contribution and crediting the remainder only with the value of the settlement—were thought justified by the greater social stake in encouraging settlements. One might well question, though, the value of a settlement that does not include all tortfeasors and preclude further litigation. From an equitable point of view, would it not be fairer to place the risk either on the settling tortfeasor (as under English law where he remains liable to contribution for the full amount of his share)[144] or on the plaintiff who is free to accept or reject the settlement offer? The plaintiffs' bar is in this instance a staunch defender of the status quo, which promotes a quick settlement with defendants of limited means, typically for the value of a minimum insurance policy. The least that could be demanded was that a settlement, to be dispositive between co-tortfeasors, should be in 'good faith' *vis-à-vis* the others, i.e. should reflect the presumptive share of the settlor's responsibility and resources. However, even this was long denied, reducing the 'good faith' requirement to a dead letter until a recent decision, much to plaintiffs' consternation, insisted on this minimal safeguard.[145]

[144] *Harper* v. *Gray*, [1985] 1 WLR 1196.
[145] *Tech-Bilt* v. *Woodward Clyde*, 38 Cal. 3d 488, 698 P. 2d 159 (1985).

Plaintiffs' stake in the existing rule consists not only in arming them with the means to sustain protracted litigation against the remainder. More important, the very threat of such a settlement acts as a potent inducement for deep pocket defendants to agree to a settlement themselves. That by itself might be considered desirable, even if it can often be criticized as mild blackmail. Less attractive is the common practice of making the release conditional on an undertaking by the settlor to cooperate with the plaintiff in his claim against the co-tortfeasors. In its most unethical form, the so-called 'Mary Carter' agreement[146] provides that the amount due under the settlement will decrease on a sliding scale if the plaintiff eventually recovers a judgment against the others above a certain threshold. That such agreements have not been flatly outlawed as contrary to public policy says much for the influence of the plaintiffs' bar in legislatures, courts, and general bar associations. The most that has been done to minimize this abuse is, in some jurisdictions, to require that such agreements be disclosed to the jury, presumably to enable them to be on guard against perjured testimony.[147]

Given these evils, the Uniform Law Commissioners have now abandoned the older model in favour of a rule deducting the whole of the settlor's presumptive share from the damages recoverable against the others.[148] This would be another important step towards the ideal of limiting liability to each party's share of fault. But unlike modification of the 'joint and several liability' rule, it has no natural lobby among traditional organized groups, including insurers. Defence attorneys and citizen groups dedicated to law reform have to be counted on in this instance to overcome the entrenched interests of the plaintiffs' bar.

[146] See Comment, 'Mary Carter Agreements: Unfair and Unnecessary', 32 *Sw. L. J.* 779 (1978).

[147] e.g. Cal. Code Civ. Proc. § 877.5.

[148] Uniform Comparative Fault Act § 6.

6

LEGAL COSTS

1. THE FEE-SHIFTING RULE

The basic rule prevailing in most countries outside the United States is that the losing litigant must pay not only his own legal costs but also those of the winner, and that such costs include the winner's attorney's fee as well as court fees.[1] In Europe this rule can be traced back to Roman origins in the later Empire,[2] but is shared also by English law, where its origins are more obscure.[3] In detail the rules vary considerably from one country to another, on such aspects as grants of discretion, exceptions on grounds of fraud, vexation, delay, or other misconduct, or on grounds of poverty. In some countries fees are officially regulated, in others like England they are freely negotiated between lawyer and client, subject however to taxation by a court official in claims for reimbursement from the losing party.[4]

Although, or perhaps because, the basic pattern of fee-shifting has been entrenched for so long, it has been rarely

[1] See Pfennigstorf, 'The European Experience with Attorney Fee Shifting', 47 *Law & Contemp. Prob.* 37 (1984) for a survey and analysis, including documents, of 12 countries.

[2] The oft-cited phrase *victum victori esse condemnandum* goes back to Justinian's Codex 3.1.13.6.

[3] See Goodhart, 'Costs', 38 *Yale L. J.* 849 (1929), reprinted in *Essays in Jurisprudence and Common Law*, ch. 10 (1931). The rule is found as early as the Statute of Gloucester (1275).

[4] RSC Ord. 62, Rule 28(2) since 1986, permits recovery of all costs 'reasonably incurred' (formerly 'necessary or proper'). Even so 'party and party' costs often fall short of 'solicitor and own client' costs. To that extent, the winner has to bear part of his attorney fees himself just as he would under the American rule. See Jackson, *The Machinery of Justice in England* (6th edn. 1972) 419–21. So a successful plaintiff might also if the defendant has no assets, even if the latter is legally aided. See *Royal Commission on Legal Services* § 13.60 and .61 (1979, Cmnd. 7648).

questioned. Theoretical justifications are not entirely uniform and are complicated by a perceptible trend in modern times to make more allowance for individualized factors, including encouraging some types of litigation and discouraging others for reasons of social policy. The most popular general theories for shifting the loss to the loser are to make the winner 'whole' and, to a lesser extent, to punish wrongdoers. The first expresses the idea that one who had to resort to litigation for asserting his legal rights, whether as plaintiff or defendant, should not be out of pocket but compensated in full. The second asserts that a litigant is blameworthy for coming to court with a losing case, all the more if he abused the judicial system by litigating from improper motives. Notably, speculations about the would-be deterrent effect of the fee-shifting rule on the volume of litigation have not played a major role. This can be explained both by the fact that the American model, to be noted presently, has never been considered a serious alternative and by the fact that deterrence of unnecessary litigation is not regarded as a significant support of the European rule. On the other hand, if access to justice should deserve greater support in consonance with modern notions of equality, it can be better accomplished by Legal Aid or Legal Expense Insurance.[5]

2. THE AMERICAN RULE

The 'American Rule',[6] under which each litigant bears his own attorney's fee and does not look to the loser for

[5] On legal aid see Cappelletti and Garth (eds.), *Access to Justice* (1978); Jackson, *Machinery of Justice in England* (7th edn. 1977) 439–50; on legal expense insurance Pfennigstorf, *Legal Expense Insurance: The European Experience in Financing Legal Services* (1975).

[6] This name, attributed to Goodhart (*supra*), became officially accredited in *Alyeska Pipeline Co.* v. *Wilderness Society*, 421 US 240, 270 (1975). The only American state to adopt the European fee-shifting rule in favour of the prevailing party is Alaska (allowing however contingent fees): see Note, 'Award of Attorney's Fees in Alaska', 4 *UCLA-Alaska Law Rev.* 129 (1974). On the other hand, the only other major country to follow the American rule is Japan, with the significant exception of fee-shifting in favour of successful tort plaintiffs. See Kojima and Taniguchi, 'Access to Justice in Japan', in Cappelletti and Garth (eds.), *Access to Justice*, vol. i, at 704–5 (1978).

reimbursement, emerged in a somewhat equivocal manner in the first half of the nineteenth century.[7] During the Colonial period, and indeed for a considerable time thereafter, attorney fees were regulated by statutes. Their purpose was to prescribe both the fees a lawyer could charge his client and those that could be reclaimed from the losing party. This practice therefore rather resembled that prevailing in England except that there scales were set not by statute but by custom. Gradually, however, lawyers began to charge higher fees than prescribed, often at first disguised as gifts, but increasingly evolving into an open practice. Thus a gap arose between fees charged to a client and those recoverable from an adversary as the official fee schedules failed to keep pace, and a short-lived practice of awarding legal fees as damages came to be disallowed. The American rule, to quote a historian, 'thus emerged from a rough compromise. Lawyers gained the right to collect large fees from their clients, while restrictions for cost recovery remained as a symbolic vestige of the old regulatory approach.'[8] This understanding was confirmed in 1848 by the Field Code of Civil Procedure in New York[9] and in 1853 by a federal statute enacting a uniform rule for all federal courts.[10]

But perhaps because of its equivocal nature and origin, the rule was accepted without attempt at explanation. That it represented legislative policy was considered sufficient, in the light of the prevailing 'formalistic'[11] legal culture. A new voice heard in its defence was that of corporate interests who found themselves increasingly a target of litigation and successfully branded fee-shifting as confiscatory. The latter part of the nineteenth century became the heyday of the American rule. Towards its end, however, much as the courts were supporting it, legislation began its inroads.

The predominant legislative purpose behind this innovative

---

[7] See Leubsdorf, 'Toward a History of the American Rule on Attorney Fee Recovery', 47 *Law & Contemp. Prob.* 9 (1984).

[8] Ibid., at 16.

[9] 1848 *NY Laws* 258.

[10] Now 28 USC §§ 1920-4.

[11] See generally White, *Tort Law in America* 31-56 (1980); Horwitz, *The Transformation of American Law, 1780 to 1860* 253-66 (1977).

intervention was to promote selective social policies by encouraging private law enforcement with a reward of attorney
fees. Examples were the Voting Rights legislation of 1870, the
Interstate Commerce Act of 1887 and the Sherman Anti-Trust
Act of 1890. This trend has sharply accelerated in recent
decades, at both the federal and the state level, to the point
where the exceptions nowadays seem almost to overwhelm the
principal rule.[12] Most marked has been its incidence in public
interest litigation.[13] For example, the right to recover attorney
fees has now been extended to all Civil Rights actions.[14]
Under numerous other statutes the government now even
'subsidizes' litigation by private individuals or groups ('citizen
suits'), who need not have suffered personal harm, against
its own agencies with respect to the implementation of
administrative policy.[15] In California, a super-activist Supreme

[12] Federal statutes are listed in App. to Justice Brennan's dissenting opin-
ion in *Marek* v. *Chesny*, 105 S. Ct. 3012, 3036-9 (1985). Note, 'State Attorney
Fee Shifting Statutes: Are We Quietly Repealing the American Rule?', 47
Law & Contemp. Prob. 321 (1984) collects information and analyses state laws,
in the aggregate no less than 1974, ranging from 146 in California to 2 in
North Carolina, and revealing a dramatic rise since the 1960s. Criteria
for fixing fees are discussed by Berger, 'Compensation Formulas for Court
Awarded Attorney Fees', 47 *Law & Contemp. Prob.* 249 (1984); id., 'Court
Awarded Attorneys' Fees: What is "Reasonable"?' 126 *U. Pa. L. Rev.* 281
(1977).
[13] See Percival and Miller, 'The Role of Attorney Fee Shifting in Public
Interest Litigation', 47 *Law & Contemp. Prob.* 233 (1984).
[14] Civil Rights Attorney's Fees Awards Act of 1976, 42 USC § 1988. See
Note, 'Promoting the Vindication of Civil Rights Through the Attorney's
Fees Awards Act', 80 *Colum. L. Rev.* 346 (1980). The fee need not be proportioned to the damages awarded (as would be contingent fees), since the
Act was passed precisely because the private market was inadequate: *City of
Riverside* v. *Rivera*, 106 S. Ct. 2686 (1986). This does not preclude contingent
fee contracts, but there is no unanimity on what weight to give to them in
determining the market rate for the services: see *Hamner* v. *Rios* (9th Cir.
1985) 769 F. 2d 1404; *Kirchoff* v. *Flynn* (7th Cir. 1986) 786 F. 2d 320; 76 ALR
Fed. 347 (1986) collecting the cases. See generally, Breger, 'Compensation
Formulas for Court Awarded Attorney Fees', 47 *Law & Contemp. Prob.* 249
(1984).
[15] See Fein, 'Citizen Suit Attorney Fee Shifting Awards: A Critical Examination of Government-"Subsidized" Litigation', 47 *Law & Contemp. Prob.*
211 (1984). Compendiously the Equal Access to Justice Act (Pub. L. No. 96-
481 §§ 203-4 (1980)) authorizes attorney fees whenever the US government's
position in litigation or agency proceedings involving individuals or small

Court exercised a traditionally legislative role by creating out of whole cloth a doctrine of 'private attorney-general' to support award of legal fees in suits by public interest litigants seeking to vindicate 'strong or societally important public policy'.[16] In previous proceedings, the successful plaintiffs had gained a ruling that the California public school financing system violated the equal rights provisions of the State Constitution.[17] A statute codifying the private attorney-general doctrine contemporaneously extended it to vindication of statutory as well as constitutional rights.[18]

Two judicially created instances of fee-shifting in the public interest are the 'obduracy' and 'common fund' exceptions. The first, in the interest of judicial efficiency, permits a successful litigant to recover attorney fees from an adversary guilty of obdurate conduct prior to or during the trial.[19] The second entitles lawyers who have been instrumental in the creation or preservation of a fund for a group of beneficiaries to reward themselves for their services out of that fund. This theory, which offended the *laissez-faire* underpinning of the American rule no more than do the ubiquitous fee clauses in contracts, was sanctioned as early as the 1880s.[20] It came to play an important role in derivative suits by minority shareholders against corporate management,[21] and in class

---

businesses is not 'substantially justified'. These statutes negated *Alyeska Pipe Line Co.* v. *Wilderness Society, supra*. Fees may be awarded so long as a plaintiff has not lost on *all* issues: *Ruckelshaus* v. *Sierra Club*, 463 US 680, 684 (1983).

[16] *Serrano* v. *Priest (Serrano III)*, 20 Cal. 3d 25, 569 P. 2d 1303 (1977). The doctrine had been expressly repudiated by the US Supreme Court as 'erroneous' in the absence of statutory authorization. *Alyeska, supra*, at 270, n. 46.

[17] *Serrano* v. *Priest (Serrano II)*, 18 Cal. 3d 728, 557 P. 2d 929 (1976).

[18] Cal. Code Civ. Proc. §1021.5. See *Woodland Hills Residents Assn.* v. *Los Angeles*, 23 Cal. 3d 917, 593 P. 2d 200 (1979), reconciling the statute with *Serrano III*.

[19] See annot. 49 ALR 4th 825 (1986).

[20] *Trustees* v. *Greenough*, 105 US 527 (1881). See generally, Dawson, 'Lawyers and Involuntary Clients: Attorney Fees from Funds', 87 *Harv. L. Rev.* 1597 (1974); id., 'Lawyers and Involuntary Clients in Public Interest Litigation', 88 *Harv. L. Rev.* 849 (1975).

[21] See Hornstein, 'The Counsel Fee in Stockholder's Derivative Suits', 39 *Colum. L. Rev.* 784 (1939).

actions generally as an adjunct to law enforcement through mass litigation.[22]

Notably, the fee-shifting in all these, mainly legislative, instances is one-sided, permitting recovery of fees by plaintiffs, but not by defendants. The more conventional judiciallycreated exceptions to the American rule, designed to promote the integrity of the judicial system, as in cases of vexatious litigation or abusive delay, operated both ways, usually to discourage litigation. By contrast, the innovative use of fee-shifting under modern legislation mostly[23] serves to promote litigation by creating an incentive for vindicating public rights or the public interest.[24] But lawyers engaged in traditional actions for tort or breach of contract do not need these innovative aids because they serve clients who are, by and large, able to pay for the increasingly expensive services. It is in these areas that the American rule has been least questioned.

2.1 Evaluation

Justifications for the American rule long seemed to be viewed as almost superfluous; it had come to be regarded as the natural thing. Such views as were occasionally expressed were often contradictory and always impressionistic. 'The first commentators tended to view litigation as an evil, but disagreed as to whether the American rule encouraged or discouraged it. Later authors emphasized the encouragement of worthy litigants, but were equally unable to agree as to whether the American rule promoted or obstructed this goal.'[25]

Fairly typical of the prevalent intuitive generalizations is the following comment by the US Supreme Court:

[22] See Kalven and Rosenfield, 'The Contemporary Function of the Class Suit', 8 *U. Chi. L. Rev.* 684, 715-16 (1941).

[23] Far less common is statutory fee-shifting in favour of the prevailing party, whether plaintiff or defendant, as under the Copyright and Patent Acts. An exceptional use of this method for the purpose of *discouraging* unreasonable litigation was contained in the Florida Medical Malpractice Act of 1977 to deal with the medical insurance crisis, but later superseded by the more conventional caps on contingent fees: *supra*, ch. 3, at n. 36.

[24] See Zemans, 'Fee Shifting and the Implementation of Public Policy', 47 *Law & Contemp. Prob.* 187 (1984).

[25] Leubsdoff, *supra*, at 28.

In support of the American rule, it has been argued that since litigation is at best uncertain one should not be penalized for defending or prosecuting a law suit, and that the poor might be unjustly discouraged from instituting actions to vindicate their rights if the penalty for losing included the fees for their opponents' counsel. . . . Also, the time, expense, and difficulties of proof inherent in litigating the question of what constitutes reasonable attorney's fees would pose substantial burdens for judicial administration.[26]

Recent studies, some buttressed by economic analysis, have pursued the effect of alternative fee rules on litigation behaviour with more theoretical rigour.[27] As a result it has become evident that the picture is much more complex than the impressionistic generalizations had suggested; indeed, that it is beset with so many contingencies as virtually to stultify confident predictions.[28]

Take the effect of alternative fee rules on the parties' decision to litigate. In the United States it is widely believed, as the above-cited statement by the Supreme Court testifies, that the English rule unjustly discourages the poor from pursuing claims and is therefore undemocratic. But many years ago, Goodhart[29] challenged this hypothesis by arguing that the poor man is at a disadvantage only if he has to give security in advance (as in India[30] but not ordinarily in England), since lack of funds prevents losing the action from being a deterrent. On the contrary, it is the wealthy defendant who is at a disadvantage because he has to pay the plaintiff's fees if he loses and cannot collect his own if he wins. Moreover, a wealthy American defendant can tire out his poor adversary

---

[26] *Fleischmann Corp.* v. *Maier Brewing Co.*, 386 US 714, 718 (1967).

[27] See Rowe, 'Predicting the Effects of Attorney Fee Shifting', 47 *Law & Contemp. Prob.* 139 (1984) who surveys the literature and considerably enhances our understanding.

[28] 'With a two-way rule there are so many cross-cutting effects and factors—encouragement from possible recovery of one's own fees, discouragement from possible liability for an adversary's fees, the presence or absence of risk aversion—that no *general* prediction of the relative overall effects of the American and English rules on the likelihood that prospective plaintiffs will pursue claims seems possible.' (Rowe, *supra*, at 147.)

[29] *Supra*, n. 3, at 874–6.

[30] Under the Court-Fees Act (VII of 1870) close on 10% *ad valorem* fees are exacted.

or force him to accept an unfair settlement, because he does not face the risk of incurring heavy extra costs.

What is the effect on the litigant of modest means? The English rule has both an incentive and a disincentive lacking in the American rule: it offers the possibility of undiluted recovery or reimbursement for a successful defence, but the risk of loss is considerably greater because of potential liability for an adversary's legal fees.[31] The net effect of this combination is uncertain, but on balance seems more likely to discourage the pressing of claims and defences by such litigants, because risk-aversion is probably strong among middle income would-be litigants, whereas wealthy and frequent litigants tend to be risk-neutral.

The effect of fee-shifting on settlement prospects is perhaps more predictable. The prospects of settlement depend on the 'bargaining span' between the parties.[32] Because that span is likely to be wider as a result of injecting the uncertainty of fee-shifting, the American rule would tend to improve the chances of settlement. The typical risk-aversion by persons of modest means would reinforce that pull: increasing the gap will also feed their anxiety. Posner also speculates that this may provide a clue as to why the fee-shifting rule is followed in England but not in the United States.[33] The more rigid adherence to *stare decisis* and the absence of juries in England make litigation outcomes more predictable and mistaken prediction therefore more avoidable. But the closer the judicial system approaches randomness, the less economic purpose is served by penalizing mistaken predictions. One might add also that the closer the issue (as in most appeals), the more questionable becomes the assumption, so commonly held in England, that reimbursing the prevailing party is an equitable imperative.

So far the discussion assumed a choice between only two alternatives, the two-way fee-shifting rule and the so-called American rule of 'each litigant for himself'. But *en passant* we already encountered one other variant, the prevailing plaintiff's one-way fee-shifting rule which is intended to, and does, encourage him to press claims without any offsetting risk of loss.

[31] Rowe, *supra*, at 153.
[32] See ch. 5, at n. 101.

Another variant of overriding significance for tort litigation, the contingent fee, now calls for particular attention.

3. THE CONTINGENT FEE

The plaintiff's attorney fees in American tort litigation are almost invariably fixed on a contingency basis.[34] If the plaintiff wins, his attorney will be paid, typically in proportion to the award recovered, say one-third; if the plaintiff loses, his attorney will be paid nothing. In England and most other countries such contingency fees are regarded as unethical or even illegal, constituting champerty.[35] The reason for this disdain is largely historical. Speculation in litigation was condemned both as an evil practised principally by powerful magnates in the medieval period as an 'engine of oppression',[36] and as discreditable like all lottery, usury, or other forms of gambling. It was particularly unbecoming of lawyers, who, in the ancient Roman tradition, regarded themselves as gentlemen and public champions, not businessmen, and who—far from trading in their clients' fortunes—could not even sue for their fees.[37]

This high-minded ethos was shared by the American legal profession until well into the nineteenth century. But as fee regulation began to crumble not long after Independence, contingency fees made their appearance in company with other freely negotiated fee arrangements between lawyer and

---

[33] Posner, *Economic Analysis of Law* (2nd edn. 1977) 452–3.

[34] See generally, MacKinnon, *Contingent Fees for Legal Services (A Report of the American Bar Foundation)* (1964).

[35] Maintenance and champerty were decriminalized in England by the Criminal Law Act 1967, but such agreements remain illegal and unenforceable: Solicitors Act 1974 s. 59(2)(b); and see *Wallersteiner* v. *Moir (No. 2)*, [1975] QB 373 (C.A.); *infra*, nn. 71–6. Generally, Zander, *Lawyers and the Public Interest* 115–20 (1968). Germany: BGHZ 22, 162 (1956) and BGHZ 44, 183 (1965) addressed the question whether contingent fee contracts between American lawyers and German clients, governed by American law, were *contra bonos mores* and hence unenforceable under EGBGB § 30 or incompatible with domestic German laws.

[36] Blackstone, *Commentaries on the Laws of England* iv, 135.

[37] In England, barristers to this day cannot sue for their fees, though this tradition is no longer universally shared in the Commonwealth.

client.[38] Gradually they came to be associated primarily with employers' liability cases, typically on behalf of poor clients. The upper echelons of the bar, on the other hand, viewed this practice with scorn. Thomas Cooley, judge, professor, and author of prominent books on torts and constitutional law,[39] thought it was no better than a 'lottery ticket' that 'brought the jury system into contempt' and helped to create 'a feeling of antagonism between aggregated capital on the one side and the community in general on the other'.[40] All the same, by 1881 the contingent fee was said to be an 'all but universal custom of the profession' despite continuing protests from the 'elder lights of the bar'.[41] Although in the course of time, workers' compensation replaced the contingent fee with a regulated fee rate in cases of employers' liability, the spectacular latter-day rise in the volume of tort litigation and in the size of awards inevitably came to be reflected in a corresponding rise in the prestige of the trial bar and of its escutcheon, the contingent fee.[42] Indeed, it has spread from tort and civil rights actions to eminent domain (condemnation of property for public purposes), bankruptcy, debt collection, even corporate take-overs, and contingency bonuses have become common in judicial awards of fees in class actions and under other fee-shifting statutes.[43]

[38] Leubsdorf, *supra*, at 16, cites decisions of 1835 and 1840, as well as *Daniel Webster's Papers*, adding that 'some lawyers had introduced contingent fee devices well before this, perhaps to evade legislative fee regulations by acquiring part of the claim'.

[39] Thomas M. Cooley, 1824–98, was one of the three foundation professors at the Michigan Law School (1859) and Judge of the Supreme Court of Michigan 1864–85. His torts book was first published in 1879. See Johnson and Malone, *Dictionary of American Biography* vol iv, 392–3 (1943).

[40] Cited by Friedman, *A History of American Law* 423 (1973).

[41] A Note in 13 *Cent. L. J.* 381 (1981) sets out the solicited opinions of three worthies, all deprecating the practice.

[42] Contingent fees have long been recognized both by Canon 2 (EC 2–20) of the Model Code of Professional Responsibility and now under Rule 1.5 of the Model Rules of Professional Conduct (1983).

[43] See Leubsdorf, 'The Contingency Factor in Attorney Fee Awards', 90 *Yale L. J.* 473 (1981), who is critical of the prevailing unsystematic practice in assessing that factor. For an interesting example under the Florida Medical Liability Act see *Fla. Patient's Comp. Fund* v. *Rowe*, *supra*, n. 23; for a spectacular award in the *Agent Orange* settlement see *In re 'Agent Orange' Products Liability Litigation*, 611 F. Supp. 1223 (1985).

3.1. Primary Effects

The contingent fee does not depart from the basic opposition of the 'American rule' to attorney fee-shifting from winner to loser. Its variation from the principal rule consists only in the special arrangement between the plaintiff and his own attorney. Relieving the plaintiff from his own attorney's fee if he loses shifts the risk of loss from him to the attorney. For this he has to pay a price, namely, a premium on the fee charged to him if he is successful.

The principal justification of the contingent fee commonly averred is that it provides easier access to justice for plaintiffs irrespective of their financial resources. Our earlier discussion of the American rule emphasized the importance of risk aversion on the decision to litigate especially by persons of modest means, concluding that on balance that rule probably had a slight edge on fee-shifting in this respect.[44] The contingent fee radically alters that balance by removing from would-be plaintiffs the deterrent of losing the action. To the extent, moreover, that it has its counterpart in reduced recovery by successful plaintiffs, it reveals itself as clearly risk-averse. That may seem somewhat paradoxical in light of the American ethos for risk venture which is often contrasted with European over-concern with security.

The rule looks, of course, very different from the point of view of the attorney who is the one to bear the risk of defeat. It is he who assumes the role of entrepreneur. If he wins, he may win 'big'; if he loses, he will have wasted his time. Perhaps that prospect is a little more favourable than that facing lawyers of indigent clients in other countries, who in the absence of legal aid would in practice also be betting on the success of the claim without any corresponding premium if the bet comes off. The paramount question for the American attorney approached by a potential client is therefore whether the case is worth taking the risk. In plain language, he calculates what the case is worth, i.e. the likely amount to be recovered discounted by the probability of failure. The larger the award, the greater will be his own reward. In practical

[44] *Supra*, n. 31.

terms, the greater the injury, the more likely that the victim will find a lawyer to take on his case; indeed, the more likely he will find a proportionally skilful attorney.[45]

It would be wrong to assume that because the institution is so widely seen as promoting the welfare of the less well-off, it might be confined to that class of clients. Most attorneys in fact choose to offer their services on a contingent basis regardless of the client's financial resources presumably because of the anticipated higher rewards than those under an hourly rate.[46]

In one aspect, the attorney is engaged in 'risk-pooling'. With many small risks, the uncertainty of his income is quite low over the long run. He is, in effect, an insurer. In another aspect, the attorney is a 'gatekeeper' with an important social role of regulating litigation, as a by-product of his own selfish financial concerns. Under Legal Aid systems it falls to an administrative agency or independent professional body to screen applicants in order to determine which have a reasonable chance of success. If the would-be plaintiff does not qualify for aid, he must make his own decision whether to run the risk of litigation. In either case, the choice will tend to be conservative. By contrast, the contingent fee system calls for a free enterprise, non-bureaucratic decision by the attorney. It contains a built-in incentive to take a chance even on marginal claims, especially if the potential reward is high. The long-term effects of this incentive on the development of tort law will be examined later.

---

[45] See *Roa* v. *Lodi Medical Group*, 37 Cal. 3d 920, 935; 695 P. 2d 164, 174 (1985) where Bird, C.J. (diss.) reasoned that (unrestricted) contingent fees provide the necessary incentive for high quality legal representation, especially needed in difficult cases like medical malpractice.

[46] The ABA Model Code of Professional Responsibility EC 2-20 still provided that 'a lawyer generally should decline to accept employment on a contingent fee basis by one who is able to pay a reasonable fixed fee'. The new (1981) ABA Model Rules of Professional Conduct (Rule 1.5) contain no such qualification.

Apparently 2-6% of individual plaintiffs, 12-15% of organization plaintiffs pay at an hourly rate, 1-2% a flat fee: Kakalik and Pace, *Costs and Compensation Paid in Tort Litigation* 37, 96 (1986).

3.1.1 Criticisms

That the contingent fee has not attracted followers outside the United States and Canada,[47] indeed is condemned as unethical or *contra bonos mores*,[48] is due to a series of negative perceptions entertained by most outside observers. Some of these objections are, no doubt, influenced by tradition more than by dispassionate analysis; others are based not on ideology, but on experience with the system's operation that has prompted controls even in its fastness, the United States.

The most vocal objection is to the widespread perception of inordinate windfalls to successful trial attorneys. While contractual terms are not absolutely uniform, the average would stipulate a fee of 33⅓ per cent. Some provide a rising scale depending on the terminating stage of the litigation, from 25 per cent if the case is settled before suit is filed to 50 per cent on appeal; others provide separate scales for specified types of claim, for example, 50 per cent for medical malpractice or products liability.[49] These premiums seem to reflect higher levels of skill and effort as well as greater risk.[50]

A common complaint is that these rates are unfair.[51] As an abstract proposition such a charge cannot be sustained. For each case depends not on the rate as such, but on the particular ratio of the skill and effort invested by the lawyer to the amount eventually recovered by him. The question is, 33⅓ or 50 per cent of what? Public attention inevitably focuses on multimillion-dollar awards when the fees look excessive; indeed, they often are: the award of $10bn. to Pennzoil against Texaco for interference with contractual relations in the

---

[47] See Williston, 'The Contingent Fee in Canada', 6 *Alb. L. Rev.* 184 (1968); Kritzer, 'Fee Arrangements and Fee Shifting: Lessons from the Experience in Ontario', 47 *Law & Contemp. Prob.* 125 (1984); Grant, 'Contingency Fees', 1980 *NZLJ* 334.

[48] *Supra*, n. 35.

[49] 50% of malpractice cases are disposed of with no recovery, compared to only 20% in automobile cases. California Auditor-General, *The Medical Malpractice Insurance Crisis in California* 28 (1975). Franklin, Chanin, and Mark, 'Accidents, Money, and the Law: A Study of the Economics of Personal Injury Litigation', 61 *Colum. L. Rev.* 1, 13 (1961).

[50] See *supra*, n. 43 for judicial awards on the same model.

[51] See MacKinnon, *supra*, ch. 9 ('The Fairness of Contingent Fees').

take-over contest of Getty Oil in 1985 was reported to yield
$2bn. to the successful attorneys;[52] the attorneys for the
asbestos victims were to receive $1bn. from the Manville
Corporation's bankruptcy settlement.[53] Such cases might argue
for controls but not necessarily against contingent fees as
such.[54] The amount recovered in most tort verdicts or
settlements is of course much more modest despite the
spectacular rise of mega-awards in recent years. Indeed,
plaintiffs' and defendants' legal costs are not far apart; at least
in routine cases (traffic accidents) plaintiff attorneys would be
earning no more if paid at the hourly rate charged by their
adversaries.[55] Finally, many other legal systems tolerate fee
structures lacking any relation between work and fee, such as
charges for conveyances in England[56] and for all legal work
in several Continental countries in proportion to the 'object
in dispute'.[57]

Abuses are, of course, possible. In theory, the occasional
windfall is justified as belated compensation for the lawyer's
unsuccessful cases. Looked at this way, the successful plaintiff
pays not only his own legal costs but also those of his attorney's
unsuccessful clients. Plaintiffs, as a class, seem to prefer not
having to pay legal fees when they lose, even if as a trade-off
they have to pay more if they win. Unfortunately, this model
would get seriously adrift if lawyers confined themselves
primarily to cases with a very high return and very high

[52] *Wall St. J.*, 21 Nov. 1985, p. 2 col. 3.

[53] Ibid., 9 Dec. 1985.

[54] Illustrative is the Florida Court's decision allowing only $1.5m. as
reasonable attorney fees (against the *defendant* under the fee-shifting provision
of the Medical Recovery Act) instead of $5m., the 40% contingent fee. *Florida
Medical Center* v. *Von Stetina*, 436 So. 2d 1022 (1983).

[55] According to a valuable Wisconsin study (App. A of Kakalik and Pace,
supra), average plaintiffs' legal expenses in auto cases were $3,000 (in 1978
dollars) for 55 hours work, that for defendants $1,900 for 30 hours. Thus the
former averaged $54.5, the latter $63 per hour. The ratio of legal expenses to
total system costs is also much the same: defendants' in non-auto cases being
18%, plaintiffs' 20%, in auto cases 13% and 24%.

[56] See Jackson, *supra*, at 515-526; Zander, *supra*, ch. 9; Abel-Smith and
Stevens, *Lawyers and the Courts*, ch. 14 (1967).

[57] See Kaplan, Von Mehren, and Schaefer, 'Phases of German Civil Procedure', 71 *Harv. L. Rev.* at 1461-70 (1958); Suhr, 'Legal Fees in Germany',
Int'l Bar J. May 1979, p. 18.

prospect of success and denying their services to others. However, this is not a serious problem in fact, because the more successful trial attorneys vie for opportunities of forensic success in speculative and spectacular cases. Besides, fierce competition at the bar means that litigants are more likely to find a champion even beyond the point where the chance of success becomes so unreasonable that it would no longer be in the public interest to encourage litigation.

Complaints about abusive practices have led to some measure of regulation.[58] In general, voluntary action by the professional bodies has been unproductive, if only because in most states they lack any effective sanction, membership being voluntary.[59] Besides, maximum scales recommended by bar associations have invariably become minimum scales. In a few states, courts eventually intervened. For example, New York offers attorneys a choice between a flat fee of 33⅓ per cent or a sliding fee schedule from 50 per cent for the first $1,000 recovered down to 25 per cent for any excess over $25,000.[60] Statutes in other states have enacted specified fee limits, following the model of workers' compensation, in areas such as medical malpractice[61] or actions against the govern-

---

[58] See Report of Defense Research Committee, IAIC, 'A Study of Contingent Fees in the Prosecution of Personal Injury Claims', 33 *Ins. Couns. J.* 197 (1966) which contains a state by state tabulation.

[59] The ABA Canon 13 demands merely in general terms that the fee be 'reasonable in all the circumstances of the case, including the risk and uncertainty of the compensation'. The ABA Model Code of Professional Responsibility and the new Model Rules of Professional Conduct demand merely that the fee be reasonable, but the latter (Rule 1.5(a)) enumerates 8 factors as relevant.

[60] *Gair* v. *Peck*, 6 NY 2d 97, 160 NE 2d 43, 188 NY Supp. 2d 491 (1959); the relevant Rule IV (for First Judicial Dept.) is set out in 33 *Ins. Couns. J.* 214–17 (1966). For its history see Note, 'Lawyer's Tightrope: Use and Abuse of Fees', 41 *Corn. L. Q.* 683 (1956). Also *American Trial Lawyers* v. *New Jersey Supreme Court*, 66 NJ 258, 330 A. 2d 350 (1974). For a survey of the practice in automobile cases see Franklin *et al.*, *supra*, at 52–63.

[61] e.g. the Medical Injury Compensation Reform Act (MICRA) of 1975 in California and a good number of other states. The California statute provides a sliding scale of 40% for the first $50,000, 33⅓% for the next $50,000, 25% for the next $100,000, and 10% for any excess: Bus. & Prof. Code § 6146. This scale reduces the maximum to less than ⅓ for cases over $140,000, to only 14% at $1m.

ment;[62] in the first example with a view to lessening the cost of liability insurance, in the second perhaps to discourage speculative litigation. These statutes have been uniformly upheld against constitutional challenge on due process grounds for allegedly impairing a client's right to retain counsel.[63]

Another complaint is that the contingent fee is apt to create conflicts of interest for the attorney. At first blush, one might well hypothesize that, since his gross earnings depend on the magnitude of his client's recovery, the contingent fee gives him a direct incentive to exert himself unstintingly in the client's interest. Indeed, the contingent fee tends to a substantial degree to monitor the lawyer's work as an hourly or fixed fee could not do, given the adventitious relationship between lawyers and clients in tort cases.[64] But there is no direct relationship between amount of fee and amount of work done. Where the fee is a flat percentage rate or a rate declining with increasing recovery, the lawyer intent on maximizing his profits could conclude that it would not pay him to invest more effort, when a knowledgeable client would be prepared to pay for extra time.[65] Thus, there is much anecdotal evidence of a widespread practice to settle claims below value on a

[62] e.g. the Federal Tort Claims Act (FTCA) of 1946, 28 USC § 2678 (20% for settlement, 25% for suit. The below-market rate is justified by the certainty of collecting the judgment.) See *Whyatt* v. *US*, 783 F. 2d 45 (6th Cir. 1986) (structured settlement).

[63] See *Roa* v. *Lodi Medical Group, supra*; appeal dism. for want of substantial federal question: 106 S. Ct. 421. Additional challenges in this case were for violation of equal protection and separation of powers doctrine; the first because the limit applied only to medical malpractice actions, the second because the setting of fees allegedly belonged exclusively to the judicial branch. The last ground succeeded in *Heller* v. *Frankston*, 76 Pa. C'th 294, 464 A. 2d 581 (1983); aff'd on other grounds: 504 Pa. 528, 475 A. 2d 1291 (1984).

[64] See *Kirchoff* v. *Flynn, supra*, at 324-6 (Easterbrook, J.).

[65] See Schwartz and Mitchell, 'An Economic Analysis of the Contingent Fee in Personal Injury Litigation', 22 *Stan. L. Rev.* 1125 (1970) for a theoretical analysis. There is a corresponding incentive for the defence to wear down the contingency-fee lawyer with dilatory tactics, excessive discovery, etc. This makes extra work for him for no pay, while the defence lawyer can bill for the time involved.

mass volume scale[66] in the belief that the cost of further litigation to the lawyer would be disproportionate to any additional profit he might derive from a larger jury award. Conversely, lawyers who have many cases a year may be more willing to take a small chance of winning big than a client who is risk-averse and has all his eggs in one basket. The only restraint would be informed client control, as contemplated by the professional ethics requirement that the lawyer keep his client informed of all relevant factors and abide by the client's decision on all important aspects of the case.[67] In practice, however, many lawyers appear to give little deference to clients on such critical matters as when to settle a case and how vigorously to contest it. Yet similar practices prevail where lawyers are paid at an hourly rate and their incentive is to pad their billing.[68]

Finally, concern has been expressed about adverse effects of the contingent fee on the judicial process itself.[69] One is that it encourages excessive, unprofessional zeal in the conduct of the client's case to 'win at all costs', from improper coaching of witnesses to groundless legal arguments designed to lead the court into error. But this impression must be balanced against the availability of judicial control and disciplinary sanctions for unethical tactics and the record of great effectiveness in legitimate trial advocacy in recent decades. Another complaint is that the contingent fee encourages unmeritorious litigation, aggravating crowded dockets and delay of trials. But this is at least tempered by the in-built discipline of the

[66] The problem is accentuated by group settlement of cases, prohibited by Rule 9 of the New York *Special Rules for the Conduct of Attorneys*, cited by MacKinnon *supra*, at 199, 228.

[67] *Model Code of Professional Responsibility* EC 7–8 (non-mandatory guide); *Model Rules of Professional Conduct*, Rule 1.4 (adopted by ABA in 1983 to replace the *Model Code*). See especially Rosenthal, *Lawyer and Client: Who's in Charge* (1974, 1977).

[68] Cf. Wolfram, 'The Second Set of Players: Lawyers, Fee Shifting, and the Limits of Professional Discipline', 47 *Law & Contemp. Prob.* 293, 294–9 (1984); *Roa* v. *Lodi Medical Group, supra*, at 945–6, 695 P. 2d at 711–12 (Bird, CJ, diss.).

[69] See MacKinnon, *supra*, ch. 10 ('The Relationship between Contingent Fees and Professional Responsibility'). In California contingent fees are categorically prohibited also for lobbyists to protect the legislative process: Gvt. Code § 86205(f).

system against wasting effort on unproductive causes; indeed, hourly rate fees seem to offer greater incentive to this form of abuse. On the other hand, the contingent fee probably does encourage speculative litigation by transferring the risk of loss from the client to the lawyer, who, in a high volume business, is more likely to be risk-neutral than a one-time client. We will revert to this tendency later in more detail.

Even 'ambulance chasing' has been blamed on the contingent fee, although only the size of the fee rather than its contingency could have a causative potential. Extensive solicitation, advertising, and fee-splitting and referral are prevalent in the field of tort litigation, but these practices reflect intense competitiveness in this branch of the legal profession, abetted by occasionally spectacular rewards. In England especially, distance between barrister and client has been an inveterate and proudly defended institutional feature. This is not the place to debate the controversial division of the profession into barristers and solicitors,[70] but one (though not perhaps its most important) justification is claimed to be the preservation of the barrister's primary loyalty to the court[71] and his dissociation from the fate and fortunes of his client. In any event, solicitors also claim the same prerogatives, which are considered incompatible with contingent fees.[72] To the extent that access to justice is acknowledged at all as a valuable social goal—curiously, little more than lip-service being paid to it even by the principal legal organizations—it is piously thought to be taken care of by Legal Aid,[73] despite the Spartan test that disqualifies most of the population and other evidence that many would-be litigants are effectively debarred from legal redress.[74] Only *Justice*, an organization of reformist lawyers, dared a proposal fashioned on a modified form of contingent fees in its submission to the Royal Commission on Legal Services. It was to set up a Contingent

[70] See generally, Zander, *supra*, ch. 12, 13 (1968).

[71] 'To a certain extent every advocate is an *amicus curiae*': Law Society testimony before the Royal Commission on Legal Services, cited by Donin, 'England Looks at the Hybrid Contingent Fee System', 64 *ABAJ* 773 (1978).

[72] See *Wallersteiner* v. *Moir* (*No. 2*), *supra*.

[73] Thus Lord Denning in *Wallersteiner*, *supra*, at 395.

[74] Zander, 'Cost of Litigation: A Study in the Queen's Bench Division', *L. S. Gaz.*, 25 June 1975.

Legal Aid Fund (CLAF) which would be interposed between attorney and client and thus avoid the undesirable features of the American contingency system.[75] A committee would screen applications to decide whether there were reasonable grounds for undertaking worthy but difficult litigation. The applicant would be required to pay the fund a predetermined percentage (from 10 to 50 per cent) of any award recovered, but could choose his own lawyer who would receive his regular fees in any event from the Fund. The Royal Commission, however, turned down the proposal on a number of grounds.[76] One was that adverse selection would quickly result in a negative balance and divert funds from other provisions like Legal Aid. Second, it was thought unfair that the successful should subsidize the unsuccessful and, last, that the system would put undue pressure in favour of parties evoking strong sympathy but with small prospects of success. Clearly, the establishment remained hostile to this avenue of experimentation.

3.2 Secondary effects

The preceding discussion has been concerned primarily with the effect of fee arrangements on litigation behaviour, with special emphasis on the comparative incentives of the contingent fee on would-be plaintiffs. Less apparent perhaps, but no less significant, are the effects of the contingent fee on the development of legal doctrine. These fall into two groups: first, the effect on damages, which has been more explicit and is linked primarily to a successful plaintiff's obligation to pay his own attorney out of the award; second, the effect on substantive law in the tendency of the contingency bonus to provide an incentive to speculative litigation.

---

[75] Justice, *Lawyers and the Legal System: A Critique of Legal Services in England and Wales* (1977). See White, 'Contingent Fees: A Supplement to Legal Aid?' 41 *Mod. L. Rev.* 286 (1978); Donin, *supra*. That contingent fees are not incompatible with fee-shifting in favour of the prevailing party is demonstrated by the experience under American fee-shifting statutes (see *supra*, n. 14), and in Alaska and Canada.

[76] Ch. 16 (1979, Cmnd. 7468).

3.2.1 Damages

I propose to explore this theme by looking in turn at collateral benefits, the taxation factor, punitive damages, damages for non-pecuniary injury, and periodical payments.

3.2.1.1 Collateral benefits

Nowadays most persons injured in an accident manage to draw on private or some sort of social insurance to meet part or all of their losses. In many Western countries, ambitious social security programmes since the Second World War have progressively reduced the tort system to the role of 'junior partner' as a source of compensation for personal injuries.[77] Although in the United States federal social security is less ambitious, its shortcomings are largely made up by privately funded, voluntary benefit schemes, such as medical insurance and pensions procured either by the individual himself or by his employer.[78]

The existence of such benefits raises the question whether they should go in reduction of tort damages or be ignored as *res inter alios acta*, i.e. none of the defendant's business. The first alternative is based on the argument that damages are compensatory and should not exceed the plaintiff's net loss. Especially benefits from public funds are intended to fill needs,

[77] *Report of Royal Commission on Civil Liability for Personal Injury*, vol. i, para. 1732 (1978, Cmnd. 7054). Commonly cited as the *Pearson Report*. See also Weyers, *Unfallschäden* (1971) for detailed comparative information and evaluation.

[78] Conard, 'The Economic Treatment of Automobile Injuries', 63 *Mich. L. Rev.* 279, 288 (1964): '[T]he actual contribution of the non-tort programs is probably greater than the 50% indicated by the Philadelphia and Michigan studies.' This conclusion is confirmed by the Dept. Transportation Automobile and Insurance study, *Economic Consequences of Automobile Accident Injuries*, 44–5 (1970), according to which only about ⅓ of the recovery for personal and family losses due to serious injury or fatality was from tort, 15% from medical and automobile insurance, 14% from life insurance, 6% from collision insurance, and 24% from wage replacement sources (sick-leave, workers' compensation, etc.). The most recent study of automobile cases reported similar results: Rolph, *Automobile Accident Compensation*, vol. i, 16–17 (1985). By comparison, the *Pearson Report* (*supra*, n. 78) concluded that in England the proportion exceeded 75% (vol. i, p. 13).

not to enrich the beneficiary, all the more so when the defendant would have contributed to the fund as much as the plaintiff and therefore cannot be accused of deriving an unfair advantage from it. In any event, rather than providing a windfall for either plaintiff or defendant, the public fund is apt to assert its own interest by claiming reimbursement for past and future benefits. In almost all countries, therefore, the social security system or its analogues are entitled to sub-rogation or reimbursement from the tortfeasor.[79] Exceptional in this regard are the United Kingdom[80] and Sweden,[81] where the view is that the cost of shifting the loss from one insurance pool (e.g. motor-car liability insurance) to another (i.e. social insurance) outweighs any conceivable social or economic benefit.[82] But either way tort damages will usually be reduced by the value of the benefit.

The competing solution of ignoring collateral benefits in assessing damages has the strongest appeal in relation to benefits paid for by the plaintiff himself. The argument is that he, rather than his injurer, should be rewarded for his foresight and thrift. Accordingly, in the case of private insurance, 'double recovery' is usually condoned (except in case of indemnity insurance subject to subrogation). The same applies to gifts, with the added purpose of enabling the donee to return the gift or loan to the donor, who is typically less able to bear the cost than the tortfeasor and his distributees. Most legal systems recognize these as exceptional situations, even if they are otherwise opposed to double recovery. In the United States, however, they have been generalized into the wider principle that the defendant should not take advantage of benefits accruing to the plaintiff from any independent source whatever.

This so-called 'collateral source rule' is far more categorical

---

[79] See Fleming, *Int'l Encycl. Comparative Law*, vol. xi (Torts) ch. 11 ('Collateral Benefits') (1986).

[80] Contrary to the Beveridge Report (1944, Cmd. 6551), subrogation was purposely rejected on the recommendation of the Monckton Committee on Alternative Remedies (1946, Cmd. 6860).

[81] See Fleming, *supra*, at 53.

[82] In some other countries such shifting is reduced or altogether avoided by loss sharing or knock-for-knock agreements between insurers: see Fleming, *supra*, § 72.

and far-reaching than the ambivalent and vacillating posture of English and Commonwealth courts. English lawmakers in particular have long been uneasy about reconciling the compensatory purpose of damages with ignoring the substantial benefits to accident victims from public sources under expanding social security programmes. A statutory compromise, originally forced on the Labour Government by the trade unions in 1948, stipulated that only one half of the value of certain benefits for five years were to be taken into account in assessing damages, in recognition of the victim's own contributions.[83] But despite criticism by the Pearson Commission that social security benefits could not realistically be likened to fruits of individual thrift and that the purpose of damages was compensation of loss rather than punishment of wrongdoing,[84] the recommended deletion of this 'privilege' has little chance of implementation in the absence of substantial corresponding increases in the value of the social security benefits. The courts, for their part, have insisted that plaintiffs cannot claim for free medical services under National Health or for wages or disability pay from employers,[85] because such benefits would have precluded any loss being incurred.[86] Beyond that, the courts invoke a somewhat elusive distinction between benefits specifically intended to compensate the victim for injury and other benefits, like retirement pensions, which have a more general purpose.[87] While an earlier search for a solution in terms of 'causation' or 'remoteness' has been replaced by a more pragmatic appeal to 'justice, reasonableness, and public policy',[88] the overall picture continues to suggest a lack of basic principle, if not disorientation.

[83] Law Reform (Personal Injuries) Act 1948 s. 2(1). The relevant benefits are industrial injury, disablement, and sickness benefits.

[84] *Pearson Report, supra*, paras. 467–76 (pp. 106–8).

[85] *Parry* v. *Cleaver* [1970] AC 1; *Graham* v. *Baker* (1961) 106 CLR 340 (wages); *Evans* v. *Muller* (1983) 151 CLR 117 (unemployment pay). *Contra*: *Boarelli* v. *Flannigan* (1973) 36 DLR 3d 4 (Ont. C. A.).

[86] A principle not consistently adhered to, as by allowing damages for the value of nursing services supplied without cost by family members on the basis of 'need' rather than 'loss incurred': *Donnelly* v. *Joyce* [1974] QB 454; *Griffiths* v. *Kerkemeyer* (1977) 139 CLR 161.

[87] *Parry* v. *Cleaver, supra*; *National Insurance* v. *Espagne* (1961) 105 CLR 569.

[88] *Parry* v. *Cleaver, supra*, at 13 (Lord Reid).

The American collateral source rule suffers from no such indecision. It applies across the board to all benefits from any source other than the defendant or co-defendants, be they cash benefits from public or private sources or benefits in kind, such as free medical services. Nor does it matter whether they were funded by contributions from the plaintiff or a third party, often the plaintiff's employer. Because subrogation provisions under statutory or contractual schemes are infrequent and quite unsystematic, the collateral source rule will generally assure multiple recovery for the victim.

The traditional justification has been that these benefits were intended to aid the plaintiff, not the defendant; indeed, that to reduce the defendant's liability would blunt the law's deterrence of wrongdoing. But this rationale has lost much credibility in the course of time. For one thing, tort liability has been pushed so far beyond common-sense notions of fault that very often to label the defendant a 'wrongdoer' or 'culprit' is to engage in hyperbole. For another, the sting of personal liability is typically blunted by the fact that the cost is passed on to an impersonal pool either through insurance or in the pricing of the defendant's products.

More convincing is the plea that tort damages do not ever make the plaintiff whole. In one sense, this seeks to convey the notion that money cannot hope to repair grave personal injuries; in another sense, one germane to the present theme, it is a reminder that, even if the award purports to indemnify the plaintiff for all his loss so far as money can do this, he must still give up a good share of it to his own attorney. This link between the collateral source rule and the contingent fee is widely acknowledged. Indeed, occasionally it has furnished the express rationale of a decision. In *Helfend* v. *Southern California Rapid Transit District*[89] the defendant protested the exclusion of evidence that the plaintiff's hospital expenses had been largely reimbursed by Blue Cross Insurance, on the ground that the Government Code prohibited punitive damages against government entities. The court first rejected the premiss that the collateral source rule was punitive in nature and therefore fell under the statutory prohibition. Rather, it

[89] 2 Cal. 3d 1, 465 P. 2d 61 (1970).

identified one of its principal purposes as being to compensate the plaintiff more fully by making up for his lawyer's contingent fee.

Generally the jury is not informed that plaintiff's attorney will receive a large portion of the plaintiff's recovery in contingent fees or that personal injury damages are not taxable to the plaintiff and are normally deductible by the defendant. Hence, the defendant rarely actually receives full compensation for his injuries as computed by the jury. The collateral source rule partially serves to compensate for the attorney's share and does not actually render 'double recovery' for the plaintiff. Indeed, many jurisdictions that have abolished or limited the collateral source rule have also established a means for assessing the plaintiff's costs for counsel directly against the defendant rather than imposing the contingent fee system. In sum, the plaintiff's recovery for his medical expenses from both the tortfeasor and his medical insurance program will not usually give him 'double recovery', but partially provides a somewhat closer approximation to full compensation for his injuries.[90]

In the result, the court while acknowledging the criticism of the rule by academicians[91] and, of course, by the defence bar,[92] concluded that reforms

cannot be easily achieved through piecemeal common law development; the proposed changes, if desirable, would be more effectively accomplished through legislative reform. In any case, we believe that the judicial repeal of the collateral source rule, as applied in the present case would not be the place to begin the needed changes.[93]

The court may have had in mind the need for linked changes, previously referred to, as illustrated by workers' compensation, which typically provides for subrogation and fee-shifting with prescribed maxima. Recent tort reforms, such as those relating to medical liability, have invariably included substantial

[90] Ibid., at 12-13. For the same reason subrogation rights were denied to an insurer of medical costs, in *Frost* v. *Porter Leasing Corp.*, 386 Mass. 425, 436 NE 2d 387, 391.

[91] Citing Fleming, 'The Collateral Source Rule and Loss Allocation in Tort Law', 54 *Cal. L. Rev.* 1478 (1966).

[92] The Defense Research Institute, an arm of the defence bar, placed abolition of the collateral source rule high on its agenda of desirable reforms. See *Responsible Reform* 21 (1969).

[93] *Supra*, n. 89, at 13.

limitations of the collateral source rule.[94] But only the Danforth
Bill on products liability has linked these to a corresponding
allowance of attorney fees for a prevailing plaintiff.[95] Another
suggestion has been to trade off the collateral source rule
against abolition of contributory or comparative negligence.[96]
Inevitably, however, the link between the rule and contingency
fees ensures that any reform has to contend with bitter
opposition from the plaintiffs' bar.

3.2.1.2 Tax liability

In the modern world, where tax liability has been steadily
escalating, in some countries to more than 50 per cent of
average earnings, the tax consequences on assessment of
damages have become highly significant.

Damage awards for personal injury are not taxable in the
United States any more than in England or most other
countries.[97] For this immunity there are a number of reasons.
Foremost is the fact that the prevailing marginal income tax
rate levied on a large award representing the present value of
many years' prospective earnings, given a system of progressive
taxation, would be quite confiscatory. Besides, it is arguable
that damages, being awarded for loss of earning capacity
rather than earnings, represent replacement of a capital asset.
Next is the problem that since verdicts and settlements are
traditionally expressed in a global sum without differentiating
between damages for pecuniary and non-pecuniary injury, it
would be difficult, if not impossible, to identify that portion
representing replacement of lost future earnings which would

---

[94] This provision, far more than 'caps', has turned out to be the most
potent in reducing awards. For Medical Reform statutes see *supra*, ch. 3, at
n. 36.

[95] The Danforth Amendment No. 1951 to s. 1999 (12 May 1986).

[96] O'Connell, 'A Proposal to Abolish Contributory and Comparative
Fault, With Compensatory Savings by Also Abolishing the Collateral Source
Rule', 1979 *U. Ill. L. F.* 591.

[97] Internal Revenue Code (26 USC) § 104(a)(2). See also *Restatement
(Second) of Torts* § 914A (1977). Punitive damages, however, are taxable, and
so is income from investment of the award, unlike periodical payments under
'structured settlements'. (In 1982 the Act was amended specifically to exempt
also awards of periodical payments.) See generally Henry, 'Torts and Taxes',
23 *Houst. L. Rev.* 701 (1986).

have been taxable if they had been actually earned. Capping all is a distaste for the tax collector's taking away from a victim of misfortune what has been awarded to console him in his plight.

But should not a jury at least be informed that the award is tax-free, lest they think otherwise and increase it accordingly? The widespread opposition to making such a jury direction mandatory seems to reflect a tolerance of such possible increases in order to offset the plaintiff's attorney costs.[98] Indeed, the US Supreme Court's direction that in federal cases the jury should be so informed[99] has not been followed by state courts.[100]

More divisive is the question whether damages must be assessed on the basis of pre-tax or post-tax earnings. In England the House of Lords chose the latter standard on the ground that it would be unrealistic, given modern tax rates and tax withholding (PAYE), to pretend that a plaintiff's actual loss exceeded his after-tax pay cheque, and that to ignore the tax incidence, in combination with a tax immunity of the award itself, would overcompensate him.[101] Nor was the court impressed by the argument that tax liability was *res inter alios acta* just like (other) collateral benefits. But it is all the more notable that its strong affirmation of the compensatory rationale of tort damages on that occasion[102] has since waned considerably in dealing with collateral benefits, strictly so-called.[103] The Australian High Court eventually followed the House of Lords.[104] Not so, however, the Supreme Court of Canada,[105] which was deterred by the problematic nature

[98] See *Hooks* v. *Washington Sheraton Corp.*, 578 F. 2d 313, 318 (D. C. Cir. 1977).

[99] *Norfolk & Western Ry.* v. *Liepelt*, 444 US 490 (1980) where the jury (in a wrongful death action which excluded non-pecuniary injury) had demonstrably doubled the amount of the proven loss.

[100] See *Klawonn* v. *Mitchell*, 105 Ill. 2d 450, 475 NE 2d 857 (1985). Cases are collected in 16 ALR 4th 589, 595–605 (1982).

[101] *British Transport Commission* v. *Gourley* [1956] AC 185.

[102] See McGregor, 'Compensation versus Punishment in Damages Awards', 28 *Mod. L. Rev.* 629 (1965).

[103] *Supra*, n. 85.

[104] *Cullen* v. *Trappell* (1980) 146 CLR 1, not following *Atlas Tiles* v. *Briers* (1978) 144 CLR 202.

[105] *The Queen* v. *Jennings* [1966] SCR 532.

of conjecturing future tax rates and the possibilities of tax planning especially for high income earners, besides the more unqualified support of the collateral source rule by Canadian courts in general. The prevailing American practice also follows the gross earnings rule.[106] Again, the reasons are mainly pragmatic. Tax savings, it is said, are too speculative and tax computation too complex for submission to a jury, apart from the fact that they invite more reliance yet on expert testimony and prolong trials. Gross income is also an economically sounder standard since it makes the defendant responsible for the full cost of the accident, thereby maximizing 'general deterrence'.[107] Again, tax savings can be viewed as a collateral benefit, a matter between the taxpayer and the government and of no concern to the defendant. All-persuasive, however, is the clincher that the benefit is more than offset by the attorney's contingent fee, so that the gross earnings rule, far from conferring an unmerited windfall, merely helps to make the plaintiff whole.[108] Comparison with the English rule is therefore peculiarly poignant, for in this instance an identical premiss—to indemnify the plaintiff, no more and no less—leads to a different conclusion, due to divergent fee rules.[109]

Admittedly, the higher the plaintiff's earnings and his probable tax liability, the more this otherwise robust solution strains credibility. According to one view, therefore, tax liability might be taken into account if the plaintiff's earnings are very high.[110] Lately, the US Supreme Court has disavowed the traditional rule entirely by holding that, in actions governed by federal law, evidence is admissible to show the effect of income taxes on the plaintiff's estimated future

---

[106] Cases are collected in 16 ALR 4th 589 (1982).

[107] For the theory that internalizing the cost to the best cost-avoider promotes accident prevention through 'general deterrence' see Calabresi, *The Cost of Accidents: A Legal and Economic Analysis* (1970).

[108] e.g. *Helfend* v. *Southern California Rapid Transit Dist.*, 2 Cal. 3d 1, 12 (cited *supra*, n. 90).

[109] In *McWeeney* v. *NY, NH & H RR*, 282 F. 2d 34, 38, n. 13 (2nd Cir. 1960), Friendly, J., specifically contrasted *Gourley* on the ground that there was no jury trial in England but fee-shifting in favour of the prevailing party.

[110] *McWeeney* v. *NY, NH & H RR, supra*, at 38.

earnings.[111] Significantly, however, once again state courts with almost one voice have refused to follow the new federal rule as not binding on questions of state law.[112]

3.2.1.3 Punitive Damages

The link between contingent fees and punitive damages is twofold: first, one of the objects of punitive damages may be to indemnify the plaintiff for his legal expenses. While the universally agreed primary purpose of punitive damages is to punish and deter egregious misconduct, a widely recognized supplementary purpose is to counteract under-deterrence by compensating the plaintiff for elements of damage such as indignity, not covered or not fully covered by the ordinary compensatory award.[113] On similar reasoning some courts specifically recognize as a proper element of damages the litigation expenses incurred by the plaintiff.[114] However, Connecticut has remained the only state to measure punitive damages exclusively by such expenses.[115]

Much more significant has been the effect of punitive damages in promoting litigation. By raising the stakes, punitive damages make the pursuit of claims worthwhile or increasingly lucrative for client and attorney. American law relies much more than the law of other countries on private initiative for law enforcement. This is reflected both in the much greater support of private causes of action for damages in addition to criminal sanctions than in England,[116] and in the numerous incentives for private law enforcement under prohibitory legislation such as treble damages, ranging from anti-trust

[111] *Norfolk & W. Ry.* v. *Liepelt, supra,* a death case; understood as a message also for personal injury: *Ruff* v. *Weintraub* (1987) 105, NJ 233, 519, A 2d, 1384.

[112] e.g. *Canavin* v. *PSA,* 148 Cal. App. 3d 512 (1983).

[113] See Ellis, 'Fairness and Efficiency in the Law of Punitive Damages', 56 *S. Cal. L. Rev.* 1 (1982). In England, since *Rookes* v. *Barnard* [1964] AC 1129, such damages are classed as compensatory and designated 'aggravated damages'.

[114] See Annot. 30 ALR 3d 1443 (1970).

[115] Ibid.

[116] A revealing comparison is the different treatment of the 'negligence *per se*' or 'statutory duty' doctrine in the two countries. See Prosser and Keeton, *Torts* § 36 (5th edn. 1984); Harper, James, and Gray, *Law of Torts* § 17.6 (2nd edn. 1986).

to anti-racketeering statutes.[117] Punitive damages occupy a prominent place in this array of public policy promotion, adding under-enforcement of rights as yet another important target of this protean remedy. Its 'private attorney general' role is most evident in its traditional core area of assault, battery, and defamation, where actual injury is often difficult to prove and where the prospect of punitive damages alone provides an incentive to pursue the aggressor.[118] The same rationale explains the award of punitive damages for violation of civil rights.[119] In the last two decades, moreover, punitive damages have become the scourge of products liability and insurance claims, yielding awards against wealthy corporations of startling magnitude. This prospect justifies the investment of substantial resources by skilful attorneys in proving reprehensible misconduct, in the expectation of thereby raising many times the financial dividend of their contingent fees. These windfalls have added a new dimension to tort law.[120]

The magnitude of the problem has been disputed. Undoubtedly, the publicity accorded to the relatively few mega-awards may exaggerate its impact, especially as many of them are subsequently reduced.[121] Nationwide statistics for 1981-

[117] e.g. Sherman Antitrust Act 1890, 15 USC § 15; Racketeer Influenced and Corrupt Organizations Act (RICO), 18 USC § 1964.

[118] In the United States damages for defamation may no longer be presumed but must be proven: *Gertz* v. *Robert Welsh, Inc.* 418 US 323 (1974); *supra*, ch. 3, n. 97. Accordingly, punitive damages in cases of malice, which are constitutionally permitted, assume a more important role than in England.

[119] *Smith* v. *Wade*, 461 US 30 (1983). Substituting $1,000 nominal damages, as recommended during the Carter administration, had little appeal, precisely because trial lawyers would have lost all incentive to undertake such actions. On the other hand, punitive damages are not allowed against the United States (FTCA, 28 USC § 2674) or foreign governments (FSIA, 28 USC § 1606), nor against most state entities (e.g. Calif. Gvt. Code § 818) 'because they would fall upon the innocent taxpayers'.

[120] Perhaps indicative of this changed perception was the reversal in 1984 of the previous exemption of punitive damages from taxation under IRC § 104 (*supra* n. 99), by replacing Revenue Ruling 7545 by Ruling 84-108. See Henry, *supra*, at 737-41.

[121] In a jury study (1982-4) final payouts were 57% of awards comprising punitive damages, compared with 82% of total awards. Most of the former resulted from judicial action, not negotiation. Shanley and Peterson, *Posttrial Adjustments to Jury Awards* 36-8 (1987).

3[122] tell us that punitive damages are still most common in cases of personal violence, fraud, and false arrest, where the amounts awarded, mostly against an individual defendant, are also comparatively modest.[123] But to this 'high incidence' group must now be added insurance, and in California business contracts, bad faith cases.[124] Punitive damage awards for the preceding categories amounted to more than 20 per cent of plaintiffs' verdicts. In products liability and malpractice cases such awards were much rarer but more than half were for $1m. or more.[125] In California between 1983 and 1985, juries awarded punitive damages in well over 200 cases, totalling $450m.; in San Francisco in 9 per cent, in Oakland 12.6 per cent of all verdicts.[126] These statistics, however, do not tell the whole story. Two other significant effects must also be added to the tally, though they are less easy to document. The first is the pervasive use made of punitive damages as a bargaining counter in settlement negotiations. The second is that the admission of evidence on the defendant's wealth may affect the jury's award of compensatory damages, even if they are told to consider it only in assessing punitive damages.[127] The vigorous campaign waged against punitive damages by the defence lobby in recent years[128] is clearly based on reality, not myth.

Conditions for Imposing Punitive Damages. Although punitive (or exemplary) damages have a respectable tradition in the

---

[122] By the ABA Research Foundation. See *L. A. Daily J.*, 11 Feb. 1986, p. 1. California and Chicago awards are analysed by Peterson, Sarma, and Shanley, *supra.*

[123] It is a rare case for punitive damages to be awarded against an individual defendant outside this category. One such was an award of $50,000 to a railroad company against a drunk driver who caused a derailment: *Nat. L. J.*, 24 Feb. 1986, p. 37.

[124] *Supra*, ch. 5, at n. 131.

[125] Thus in San Francisco median awards, 1980-4, were $63,000, average awards $381,000. Peterson, Sarma, and Shanley, *supra*, at 15.

[126] *LA Times*, 3 Nov. 1985, pp. 1, 27; Peterson, Sarma, and Shanley, *supra*, at 32-5. See also the chart of 16 appellate decisions in *Devlin* v. *Kearny Mesa AMC/Jeep/Renault, Inc.*, 155 Cal. 3d 381, 393-6 (1984); augmented in 16 *CTLA Forum* 100-3 (1986).

[127] A proposed remedy would be to admit such evidence only after the jury had first found liability and entitlement to punitive damages.

[128] See *infra*, nn. 159-60.

common law,[129] the ways of England and America have long parted. The House of Lords in 1964 reinterpreted the long but thin line of English precedents as based not on exemplary but on 'aggravated' damages, i.e. to compensate the plaintiff for his own outrage and distress rather than to punish the defendant.[130] In principle, exemplary damages were justifiable only against oppressive, arbitrary, and unconstitutional acts of government servants or where the actor sought to make a profit from his tort.[131] Courts in the older Commonwealth have declined to follow this lead, in the belief that punitive damages still serve a useful deterrent against gross misconduct in general.[132] In the United States, the practice of awarding punitive damages is deeply entrenched and, far from fading, has undergone a startling rejuvenation in recent decades.[133]

The remarkable increase in popularity coincides with the rise in influence of the plaintiffs' bar, abetted by its burgeoning representation in the ranks of the judiciary especially in Democratically controlled states like California and New Jersey. Instrumental have been both relaxation in the conditions justifying awards and a remarkable increase in the size of verdicts. To these we will now turn.

Time was when punitive damages were confined to outrageous misconduct. Gradually, many courts contented themselves with the lesser requirement that the defendant merely intended to inflict injury, ending up with the even more forgiving test of 'conscious disregard of the safety of others'.

---

[129] For more modest Continental analogues, like the German *Busse*, see Stoll, 'Penal Purposes in the Law of Tort', 18 *Am. J. Comp. L.* 3 (1970).

[130] *Rookes* v. *Barnard* [1964] AC 1192.

[131] Rare examples of the latter are *Broome* v. *Cassell* [1972] AC 1136, *Riches* v. *News Group Newspapers* [1986] QB 256 (CA) (libels to increase circulation). A few statutes, like the Copyright Act, specifically authorize exemplary damages.

[132] *Australian Consolidated Press, Ltd.* v. *Uren* [1969] 1 AC 590 (Australia); *Taylor* v. *Beere* [1982] 1 NZLR 81. For Canada see Cooper-Stephenson and Saunders, *Personal Injury Damages in Canada*, ch. 13 (1981).

[133] See generally Ghiardi and Kircher, *Punitive Damages: Law and Practice*, 2 vols. (1981, looseleaf); Redden, *Punitive Damages* (1980). Only 4 states disallow punitive damages (La., Mass., Nebr., Wash.), 3 others treat them as compensatory (NH, Mich., Conn.).

Thus even California's statutory limitation of punitive damages to tort actions for 'oppression, fraud or *malice*'[134] has been diluted to include this attenuated form of misconduct, described as 'recklessness', which in practice comes close to mere aggravated negligence.[135] Two examples will illustrate this progression. In *Taylor* v. *Superior Court*[136] the California Court affirmed that the test was satisfied in *all* cases of drunken driving, reflecting as it does 'conscious disregard', which was 'conduct that may be called willful or wanton' and therefore qualifying as 'malice'. This was said notwithstanding the conventional understanding that malice called for knowledge that harm to others was substantially certain or at least highly probable, when in fact persons who drive under the influence often lack a conscious appreciation of the high risk to themselves and others.

Even more remarkable has been the manipulation of concepts in order to subject so-called 'malicious breaches of contract' to punitive damages. This development originated in efforts, already noted,[137] to force insurers to live up to their implied covenant of good faith and fair dealing in payment and settlement of claims. Such a breach, it was held, if only negligent, deserved an award of damages for mental suffering and, in case of 'oppression', of punitive damages. The problem that punitive damages are not allowed for breach of contract was overcome by simply declaring that such a 'bad faith' breach also constituted a tort.[138] This novel quasi-tort of 'malicious breach of contract' has since been extended to employment contracts as a remedy, for example, against retaliatory discharge, and more recently, to any denial of the

---

[134] Civil Code § 3294. Italics added.

[135] Even the US Supreme Court sanctioned punitive damages for 'reckless or callous' misconduct in *Smith* v. *Wade, supra*. Contrast the conservative, minority view of the Maine Court in *Tuttle* v. *Raymond*, 494 A. 2d 1353 (Me. 1985) which insisted on the malice standard, meaning ill-will to the plaintiff or outrageous conduct (more reprehensible than reckless disregard).

[136] 24 Cal. 3d 890, 598 P. 2d 854 (1979).

[137] *Supra*, ch. 5, at n. 130.

[138] *Crisci* v. *Security Insurance Co.*, 66 Cal. 2d 425, 426 P. 2d 173 (1967) (mental suffering); *Neal* v. *Farmers Insurance Exchange*, 21 Cal. 3d 910, 582 P. 2d 980 ($740,000 punitive damages).

existence of a contract.[139] This remarkable episode is surely the most blatant example of tendentious reasoning.

The size of permissible awards for punitive damages has become even more startling. Awards used to be kept in bounds, as they still are in Commonwealth jurisdictions,[140] by the requirement that they ought to be proportionate to compensatory damages awarded for the actual injury suffered[141] and by the fact that, in the traditional area of physical aggression, the defendant's typically modest resources inevitably set a low limit on what he could be expected to pay. The first requirement has, in the United States, been subordinated to the principle that in order to make the defendant's penance realistic the jury is entitled to take into consideration his financial resources.[142] Since the extension of punitive damages to products liability and malicious breach of contract, they are now available against corporate defendants with enormous assets.[143] The proclivity of modern

[139] *Seaman's Direct Buying Service* v. *Standard Oil of California*, 36 Cal. 3d 752, 686 P. 2d 1158 (1984). The theory of malicious (tort) breach of contract was emphatically rejected in *A. & E. Supply Co.* v. *Nationwide Mutual Fire Ins. Co.*, 798 F. 2d 669 (4th Cir. 1986).

[140] See *Riches* v. *News Group Newspapers, supra,* setting aside an award of 250,000 as disproportionate to £3,000 compensatory damages. In *Broome* v. *Cassell, supra,* at 1081, Lord Hailsham LC, while not agreeing with 'Lord Devlin that appellate courts might more readily interfere with jury awards of exemplary damages than in other cases, regarded it as extremely important that judges make sure in their direction that the jury is fully aware of the danger of an excessive award'; Gibbs, CJ, in *XL Petroleum* v. *Caltex Oil*, 59 ALJR 352, 358 (1985), adding, 'I respectfully agree with those pleas of moderation'. The High Court of Australia confirmed a verdict, originally for $400,000, reduced to $150,000, taking into consideration both the high-handed conduct of the defendant (trespass to land to destroy a competitor) and its wealth. In *Broome* v. *Cassell* the House of Lords upheld an award of £25,000 (£14,000 compensatory).

[141] See Peterson, Sarma, and Shanley, *supra,* at 56-64, for a study of the relationship of punitive to compensatory damages in California 1980-4. In business/contract cases the median ratio was 1.6, the top 25% ratio 4.5, in intentional torts 1.2 and 3.0, in personal injury 0.5 and 2.1.

[142] See annot. 'Sufficiency of Showing of Actual Damages to Support Award of Punitive Damages', 40 ALR 4th 11 (1985).

[143] This is also permissible in Commonwealth countries, as demonstrated by *XL Petroleum* v. *Caltex Oil, supra,* where the original award of $400,000 represented 1% of the defendant's after-tax profits (see at 357, 362). Significantly, however, a majority of the Australian High Court nevertheless

juries to vent their individual or collective resentment against
such deep pocket defendants has already been the subject of
comment in a previous chapter.[144] Thus in one celebrated
case involving the notorious Pinto car, the manufacturer of
which was accused of not adequately protecting the gasoline
tank against rear collisions, the jury awarded $125m. to the
next of kin of three incinerated occupants.[145] The award, at
least as reduced by the trial judge to $3.5m., was upheld as
'not so grossly disproportionate as to raise the presumption
that it was the product of passion or prejudice', considering
the degree of reprehensibility of the defendant's conduct, its
wealth, the amount of compensatory damages, and the amount
that would serve as a deterrent.[146] This case also highlights
the arbitrariness of awards: besides the discrepancy between
$125m. and $3.5m. (a ratio of 40 : 1), the appeal court would,
no doubt, also have deferred to the trial judge if he had
reduced the award to $1m. or $5m. Not content merely with
verdicts of this magnitude, most courts have tolerated the
repeated imposition of punitive damages against the same
defendant. This has become common practice in mass-accident
litigation, not only in consecutive separate actions before
different judges but even in consolidated actions by multiple
plaintiffs against the same defendant.[147] Such 'overkill' is the
more difficult to justify when each award is based on the assets

thought the award excessive and agreed to its reduction. Proportionality was
also disregarded, compensatory damages amounting to only $5,527.

[144] *Supra*, ch. 4, at n. 36.

[145] *Grimshaw* v. *Ford Motor Co.*, 119 Cal. App. 3d 757, 174 Cal. Rptr. 348
(1981). *Chodos* v. *Insurance Co. of North America*, 126 Cal. App. 3d 86, 178 Cal.
Rptr. 831 (1981) upheld a verdict of $220,000 punitives as against $146
for material damage (fender bender) and $5,000 for emotional injury. See
generally Owen, 'Problems in Assessing Punitive Damages against Manufacturers of Defective Products', 49 *U. Chi. L. Rev.* 1 (1982).

[146] Ford's net worth in 1976 being $7.7bn. and its after-tax income
$983m., the award was only 0.005% and 0.03% respectively. On the other
hand, the original jury award was 44 times the amount of the compensatory
damages ($2.6m.). Since that was the reason for reducing the award, did not
the evidence of 'passion and prejudice' taint the whole verdict?

[147] Warning against such 'overkill' is Surrick, 'Punitive Damages and
Asbestos Litigation in Pennsylvania: Punishment or Annihilation?', 87 *Dick.
L. Rev.* 265 (1983).

of the defendant, because in the aggregate they can quickly exhaust the defendant's entire resources and drive it into bankruptcy, as happened in the asbestos affair. A rare voice of protest was sounded by the Second Circuit in *Roginsky* v. *Richardson–Merrell*,[148] warning that a previous award of punitive damages with respect to the same drug by a California Court was a sufficient penalty. But is it fair to award the first plaintiff(s) to the exclusion of later litigants? The very idea of repeated awards suggests that punitive damages are recognized at least as much as a reward for the plaintiff and his attorney as a penalty for the defendant. The uncoordinated multiple imposition of penalties on a defendant for the same transgression might also be fraught with constitutional infirmity.[149] One method occasionally sought to address this problem is for a court to certify a class action with respect to punitive damages so as to assure equitable distribution of the 'limited fund' of defendant's available assets for all members of the class.[150]

A related problem is whether the defendant's liability insurance will cover awards of punitive damages. Surely, to allow the defendant to escape personal liability would deprive the remedy of its intended deterrent function.[151] By common law or statute, insurance against liability for intentional or wilful injury is universally deemed to be against public policy.[152] Should not the same rationale apply to punitive

---

[148] 378 F. 2d 832 (2nd Cir. 1967). By Friendly, J., the other members of the court rejecting punitive damages on another ground. Evidence of prior awards has been more readily admitted to allow juries to reduce an award accordingly: *Fischer* v. *Johns–Manville Corp.*, 103 NJ 643, 512 A. 2d 466, 480 (1986); Redden, *supra*, § 4.8.

[149] For violating the sense of 'fundamental fairness' required by 'due process'. See Jeffries, 'A Comment on the Constitutionality of Punitive Damages', 72 *Va. L. Rev.* 139 (1986).

[150] But most such efforts have failed (on appeal) for a variety of reasons: *infra*, ch. 7, at n. 32.

[151] There is a way out: crediting the defendant with the amount paid by his insurer as part of his income: King, 'The Insurability of Punitive Damages: A New Solution to an Old Dilemma', 16 *Wake Forest L. Rev.* 345 (1980). Or allowing the liability insurer to recover the punitive damages from the insured tortfeasor, as in *Ambassador Ins.* v. *Montes*, 76 NJ 477, 388 A. 2d 603 (1978).

[152] e.g. California Insurance Code § 533; Civil Code § 1668.

damages?[153] That a substantial number of courts have none the less spurned this conclusion, usually on the sophistical ground that there is no reason for bailing out the insurer,[154] might suggest that they are more concerned about the disappointment this would create for the trial bar than about the implementation of public policy this remedy is supposed to promote. An alternative, more charitable, explanation might be the expectation that insurability would placate critics of the extension of punitive damages beyond actual intent to injure.

Indeed, insurance is not the only example of how redistribution of the effective incidence of the cost burden has eroded the deterrence rationale of punitive damages. The very application of punitive damages against corporate defendants, which has captured all recent attention, suggests that in many, perhaps most, instances the cost is being routinely passed on to outsiders: to customers, to suppliers, and to the fisc (damages are treated as a business expense and therefore tax-deductible[155]).[156] Even to the extent that it has to be absorbed by the defendant corporation itself, it is the innocent and powerless shareholders rather than the culpable officials who will pay the bill. Whether windfalls to plaintiffs and their attorneys at the expense of these groups can be justified overshadows contemporary efforts at reform.

Reform. Directly addressing itself to the crux of attorney fees is the abolition of punitive damages in exchange for allowance of reasonable attorney fees against the defendant.

---

[153] If this were so, one consequence of awarding punitive damages could be to deny the insured coverage not only for the punitive but also for the compensatory damages. See Clark, J., diss., in *Taylor* v. *Superior Court, supra,* n. 136, at 905. The only escape from this logic is that punitive damages are frequently awarded for recklessness, not amounting to 'willful or intentional acts': see *Ford Motor Co.* v. *Home Insurance Co.,* 116 Cal. App. 3d 374, 382-3 (1981). Cases are collected in 16 ALR 4th 11 (1982).

[154] See Prosser and Keeton, *supra* at 13. The two views are well stated in *Harrell* v. *Travelers Indemnity Co.,* 279 Ore. 199, 567 P. 2d 1013 (1977). Surprisingly, considering the California courts' pro-plaintiff bias, coverage is there denied: *Ford Motor Co.* v. *Home Insurance Co., supra.*

[155] IRS Ruling 80-211.

[156] See Launie, 'The Incidence and Burden of Punitive Damages', 53 *Ins. Couns. J.* 46 (1986).

This reform, in a few states,[157] detracts greatly from the attraction of contingent fees for the plaintiff's lawyer, but meets the argument that punitive damages would alone serve to indemnify the plaintiff for his legal costs.

Most reforms, however, fall short of complete abolition of punitive damages. Addressing the criticism that punitive damages have become excessive, several jurisdictions have either put a 'cap' on punitive damages or limited them to a specified ratio to compensatory damages.[158] Another approach is to raise the standard of proof so as to meet a possible constitutional objection that the civil standard of deciding the issue on the balance of probabilities constitutes a denial of due process.[159] A substitute would be proof by 'clear and convincing evidence', a test familiar in constitutional litigation, halfway towards the criminal standard of proof beyond all reasonable doubt.[160] This would increase the control of the trial judge, though not as much as the alternative proposal to take the assessment of punitive damages away from the jury altogether and entrust it, like sentencing of criminal defendants, to the judge.[161] Yet another proposal has been to protect a defendant from more than one award.[162]

In order to test the sincerity of the trial lawyers who are wont to defend punitive damages as a needed spur to accident prevention, a most provocative reform would be to divert punitive damages from the plaintiff to humanitarian purposes such as organizations devoted to industrial safety and accident prevention, the Red Cross, and the like.[163] Finally, with a

[157] Ga., Haw., Ido. For bad faith refusal to settle insurance claims.

[158] Mont. ($100,000), Colo. and Okla. (ratio 1 : 1), Fla. (ratio 3 : 1).

[159] The contention that awards of punitive damages should meet the standards of criminal trial (presumption of innocence, unanimous verdict) was rejected in *Toole* v. *Richardson–Merrell, Inc.*, 251 Cal. App. 2d 689 (1967).

[160] Adopted in *Linthicum* v. *Nationwide Life Ins. Co.*, 723 P. 2d 675 (Ariz. in banc, 1986); *Wangen* v. *Ford Motor Co.*, 97 Wis. 2d 260, 294 NW 2d 437 (1980). By statute in Minn., Oreg. A Colo. statute adopted 'beyond all reasonable doubt'.

[161] Model Uniform Product Liability Act § 120.

[162] Senator Kasten's bill, the proposed Uniform Products Liability Act, which has been promoted since 1982, besides adopting the stricter standard of proof originally also contained this proposal.

[163] The Polish Civil Code, Art. 448 provides a model for the Red Cross. See Wagner, *Polish Civil Law*, vol. ii (Obligations in Polish Law) 146 (1974).

view to curbing the size of awards, two trial judges recently
offered to reduce large awards if defendants undertook remedial measures.[164] This tactic would promote future accident
prevention, but reduce the incentive to litigate. Like other
proposals, it has therefore encountered strong condemnation
from the trial bar.

3.2.1.4 Non-pecuniary damages

In comparison to other countries which share similar cultural
values and living standards,[165] American awards for non-
pecuniary damages tend to be strikingly larger.[166] While this
disparity could arguably be seen to reflect greater sensitivity
to psychic and other non-material values in an affluent and
indulgent society,[167] a more realistic explanation would be

Florida's Tort Reform and Insurance Act 1986, besides limiting the availability and amount of punitive damages, directs that 60% be paid into Pub.
Med. Assistance Trust or Gen. Revenue and that attorney's fees be based on
the plaintiff's 40%.

[164] See Note, 'Remedial Activism: Judicial Bargaining With Punitive
Damage Awards', 19 Loy.-LA L. Rev. 941 (1986).

[165] See Szöllösy, 'The Standard of Compensation for Personal Injury and
Death in European Countries', 1983 Nordisk Forsikringstidsskrift 128. Gener-
ally, id., Die Berechnung des Invaliditätsschadens im Haftpflichtrecht Europäischer
Länder (1970).

[166] The evidence is mainly anecdotal. But according to one 'closed claims'
insurance study, the mark-up ratio ranged from 5.3 for litigated cases to 2/2.3
for attorney-assisted cases to 1.5/1.7 for others (O'Connell and Simon, Payment
for Pain and Suffering 16-18 (1972)). The American Insurance Association
(AIA) reported 3.4 for victims with attorneys and 2.3 for those without (ibid.,
at 16). The ratio nationwide for pain and suffering over $100,000 in 1984
was estimated at 4.41, rising to 5.06 (H. Manne (ed.), Medical Malpractice
Guidebook, 1985). For England the Pearson Commission, supra, para. 382,
ventured that non-pecuniary damages account 'for more than half of all tort
compensation for personal injury and for a particularly high proportion of
small payments'.

[167] Indeed, the very concept of compensating such injury has been
ascribed to the effete materialistic values of a capitalist society. Such damages
are not allowed either in China or the Soviet Union but have become so
entrenched in Western culture that other Eastern European socialist societies
have generally resisted conforming to that model. In the common law non-
pecuniary damages were first associated with intentional injury, just like
punitive damages, but were extended to negligent injury by the end of the
eighteenth century, although occasionally challenged in the United States as
late as the mid-nineteenth century. See O'Connell and Bailey, 'A History of
Pain and Suffering', in O'Connell and Simon, supra, 83-109.

found in the twin factors of trial by jury and an aggressive trial bar.

As previously explained,[168] American juries are apt to express their compassion for the individual victim before them without undue concern for the deep pocket of faceless defendants. Indeed, the defendant's deep pocket is seen as a special target for retribution, especially when the plaintiff's injuries are great and the defendant's culpability demonstrable. The generality of the instruction the jury receives from the judge, the absence of any tariff or standards that might be derived from prevailing practices as a guide, and the ambitious targets suggested by plaintiff's council,[169] are an open invitation to respond in a spirit of generosity without much countervailing restraint. Judicial controls are limited by a pronounced reluctance to intrude upon the traditional prerogative of the jury.

The driving force behind the dramatic award explosion in recent decades is the skill of the plaintiff's bar, motivated by the contingent fee. Symbolized by Melvin Belli's campaign for 'The More Adequate Award' in the early 1950s,[170] the trial bar has seen its task even more in enlarging awards than in expanding the area of substantive liability. Representation of a claimant by a lawyer has a galvanizing effect on insurance adjusters. Without the lawyer, the claimant will rarely be offered anything beyond medical costs and other economic losses; with a lawyer he will on average receive three or more times as much as without him.[171] If anything, this suggests the effectiveness of legal representation; the spur behind it is the lawyer's self-interest in increasing his own dividend.

[168] *Supra*, ch. 4, at n. 40.

[169] Most jurisdictions allow disclosure to the jury of the, typically grossly inflated, amount of damages claimed by the plaintiff (the so-called *ad damnum* clause of the pleadings). This is opposed by the defence bar as conveying a distorted view to the jury and public: see *Responsible Reform* 25 (1969). (By contrast, the practice in some countries of linking filing fees to the amount claimed (*supra*, n. 57) would act as a brake.) Besides, plaintiff's counsel typically make detailed submissions, aided with 'blackboard evidence', on *quantum*. A favoured method permitted in some jurisdictions, is the '*per diem*' argument (see 3 ALR 4th 940 (1981).

[170] Belli, 'The Adequate Award', 39 *Calif. L. Rev.* 1 (1951); id., *The More Adequate Award* (1952). [171] *Supra*, ch. 5, at n. 115.

Juries are also widely suspected of including an allowance for the plaintiff's attorney fees in their awards. To the extent that this is true, it results in fee-shifting, however surreptitious, from the defendant to the plaintiff as under the 'English rule'. Indeed, any official introduction of fee-shifting has been opposed on the very ground that fees are often already included in awards, with the result that the defendant would be at risk of paying twice.[172]

It is a common observation that non-pecuniary damages are in effect earmarked for the plaintiff's attorney. Even to the extent that this is true, it does not link the 'American rule' to the very recognition of non-pecuniary damages, since most legal systems of the Western world combine it with fee-shifting.[173] Clearly, such damages serve a variety of other objectives as well: the symbolic one of expressing moral outrage or giving 'satisfaction', of making up for under-assessment of material loss. The contingent fee rule, however, does create an incentive for plaintiffs' attorneys to press juries for the highest awards on account of an injury the value of which, unlike material loss, is elastic and least amenable to judicial control. It also explains, of course, the profession's vested interest and rhetoric in vehemently opposing the replacement of torts by compensation systems which are usually confined to benefits for pecuniary loss. Reformers in many common law jurisdictions have been casting around for novel curbs on the amount of awards. In this endeavour the traditional defence interests have been able to enlist also advocates of no-fault compensation who are content to sacrifice compensation for non-pecuniary injury for the sake of affording better compensation for actual pecuniary losses. In England, for example, where more than half of tort damages awarded are attributable to non-pecuniary loss,[174] the Pearson Commission was divided on whether to retain damages for pain and

---

[172] See e.g. *Wall St. J.*, 29 May 1986 for anecdotal evidence. Inclusion of attorney fees in the award constitutes jury misconduct and can justify impeachment of the verdict, as in *Krouse* v. *Graham*, 19 Cal. 3d 59, 562 P. 2d 1022 (1977). See generally, Comment, 'Impeachment of Jury Verdicts', 25 *U. Chi. L. Rev.* 360 (1958).

[173] See McGregor, *Int'l Encycl. Comp. L.* vol. xi (Torts) ch. 9, pp. 15-20 (1986).

[174] *Pearson Report*, para. 81.

suffering, as distinct from loss of 'faculty',[175] but recommended against non-pecuniary damages for the first three months.[176] About half the Commission members also favoured a ceiling of five times average industrial earnings (about £20,000 in 1977).[177] In Canada, the Supreme Court took it upon itself in 1978 to place a $100,000 'cap' on all non-pecuniary damages,[178] followed by the Irish Court's cap of £150,000 in 1984.[179] In the United States, automobile no-fault plans offer benefits only for medical costs and loss of earnings, and in several jurisdictions exclude tort claims altogether for less severe injuries.[180] 'Caps' on non-economic damages have also been prominent targets of most other tort reform proposals.[181] These, as already noted, have met with some success in regard to medical liability claims and are currently strongly promoted in regard to products liability and as a means to modify the 'joint and several liability' rule.[182] Not surprisingly they have been bitterly, and before many legislatures successfully, opposed by the organized trial bar.

Professor O'Connell, an untiring advocate of tort reform, has also made a proposal to abolish damages for pain and suffering in return for payment of the plaintiff's attorney fees.[183] He points to the common assumption that a substantial

[175] Ibid., paras. 380–1.

[176] Ibid., para. 388.

[177] Ibid., para. 391. In 1977 the highest award to date had been £243,000: ibid., para. 82.

[178] *Andrews* v. *Grand & Toy Alba.* [1978] 2 SCR 229 ($100,000 for quadriplegic; since adjusted for inflation: *Lewis* v. *Todd* [1980] 2 SCR 694 ($195,000)).

[179] *Sinnott* v. *Quinnsworth* [1984] ILRM 523, 532.

[180] *Supra,* ch. 5, at n. 95.

[181] The effect of a cap of $100,000 in Florida is analysed on the basis of statistics in *Medical Malpractice Policy Guidebook, supra,* at 132–41, concluding that it would affect primarily cases with large pecuniary damages (over $100,000). The Florida measure was held unconstitutional for dealing with more than one subject: *Evans* v. *Firestone,* 457 So. 2d 1341 (1984).

[182] *Supra,* ch. 3, at n. 80. In California, overreaching the legislature, a popular initiative in 1986 limited recovery of non-economic loss to the proportion of each individual defendant's fault.

[183] 'A Proposal to Abolish Defendants' Payment for Pain and Suffering in Return for Payment of Claimants' Attorneys' Fees', 1981 *U. Ill. L. F.* 333. For an earlier proposal by O'Connell to abolish the collateral source rule see *supra,* n. 96.

portion of the plaintiff's award of non-pecuniary damages is in effect absorbed by his legal expenses, so that the suggested trade-off will not prejudice either the plaintiff's or his attorney's legitimate claims. On the other hand, it would discourage the prevalent practice of padding claims and exploiting the nuisance value of trivial injuries, and avoid the wasteful and perplexing inquiry into the dollar value of the alleged injury. But since these disadvantages of the present system tend to enhance the compensation payout, trial attorneys have a vested interest in its continuance and are bound to resist reform.

Emotional distress. The successes of the plaintiffs' bar in promoting non-pecuniary damages have not been confined to the traditional area of physical injury. Several other extensions deserve notice.

First is the extension of such awards to claims for wrongful death. The narrow construction of wrongful death statutes, originally adopted in the United States as in England, limited awards to actual pecuniary losses of the survivors, excluding grief and sorrow as well as loss of companionship and society.[184] For a long time, Western states of America were alone in permitting recovery for the latter type of loss, but have been joined more recently by a growing number of jurisdictions, now amounting to 'a clear majority'.[185] The US Supreme Court considered this trend so pervasive that it adopted this legislative pattern for wrongful death on territorial waters in the exercise of its admiralty jurisdiction.[186] Nowhere is the discrepancy in the size of English and American awards more pronounced than in this context.[187] Especially in the case of

[184] See generally Prosser and Keeton, *supra*, at 951-2; Harper, James, and Gray, *supra*, § 25.14.

[185] *Sea-Land Services, Inc.* v. *Gaudet* 414 US 573, 587 (1974).

[186] Ibid. Curiously, the Court declined to adopt the closer statutory analogy of the Death on the High Seas Act (DHSA) which excludes such claims. In *Moragne* v. *States Marine Lines, Inc.*, 398 US 375 (1971), under which the *Gaudet* claim was brought, the Court had adopted a civilian approach by following the pattern of state wrongful death acts as an analogical premiss for creating a judge-made right for death on territorial waters of the US.

[187] This was dramatically illustrated in the course of the Turkish Airline Crash litigation. In the *Kween* case plaintiffs had originally been offered the equivalent of $3,400 in London; a Los Angeles jury awarded $1.5m. See Johnston, *The Last Nine Minutes*, ch. 25 (1976).

fatal injuries to a child, American parents are now able to finesse the traditional 'child-labour' standard and recover substantial awards for their loss of companionship, just as can spouses and children who have lost in the deceased more than a bread-winner.[188]

Traditional common-law support of relational claims in case of non-fatal injury was even less generous. Only a husband had a claim for loss of consortium, not a wife, parent, or child. But here too, many courts have been pressed to extend reciprocal rights which are almost always principally for non-pecuniary elements. Take the case first of the injured husband. He will be the one to claim for his lost medical expenses and earnings, aside from his own pain and suffering. This would leave the wife only a claim for her non-pecuniary injury—her loss of society and companionship, including sex relations. Rather than abolish the husband's action, most courts have instead extended it to the wife. This has occurred despite the archaic origin of the action and the problematic nature of monetary damages for so private, impalpable, and incalculable a loss.[189] In the case of the parental and filial relationship there was no similar precedent. All the same, many courts adopted the spousal model also for these relations.[190] Somewhat surprisingly, the California Court balked, invoking the whole litany of arguments against non-pecuniary damages that they had consistently contemned in the past.[191] The resulting distinction between injury and death, which was challenged on constitutional grounds,[192] was explained as resting on the rational basis that, in the case of death, there would otherwise have been no sanction against the tortfeasor

[188] In England the Fatal Accidents Act of 1976, following Irish and Australian legislation inspired by the Scottish *solatium*, authorized awards of 3,500 for bereavement.

[189] See generally Prosser and Keeton, *supra*, at 931-2. *Ossenfort* v. *Associated Milk Producers, Inc.*, 254 NW 2d 672 (Minn. 1977) upheld an award of $500,000 to the wife of a husband who suffered devastating injuries.

[190] e.g. *Theama* v. *City of Kenosha*, 117 Wis. 2d 508, 344 NW 2d 513 (1984); *Ueland* v. *Pengo Hydra-Pull Corp.*, 103 Wash. 2d 131, 691 P. 2d 190 (1984).

[191] Including the fact that in the case in question, the injured father had no less than 6 children. (The dissent pointed to the statistical average of 2.1 children.) *Borer* v. *American Airlines*, 19 Cal. 3d 441, 563 P. 2d 858 (1977).

[192] For violation of 'equal protection' see *supra*, ch. 3, at n. 3.

and that this provided the only means by which the family unit could recover compensation for loss of parental care and services, whereas in case of injury 'the tangible aspects of the child's loss can be compensated in the parent's *own* cause of action'.[193]

But this minor setback was made up by major extensions of recovery for mental disturbance.[194] American courts had resisted such claims much longer than the English. Most insisted on 'impact', in deference to pervasive fear of fraudulent claims and the presumed inability of juries to cope with conflicting and meretricious psychiatric evidence. Even after this barrier fell in the 1950s and 1960s, plaintiffs still faced a second hurdle in the requirement that they must have suffered shock as a result of fearing for their own safety rather than that of a third person, or, what usually amounts to the same thing, must have been within the area of physical danger from impact. Eventually California proudly took the lead in 1968 in the celebrated case of *Dillon* v. *Legg*[195] of allowing recovery to a mother who had witnessed injury to her child from a position of personal safety. Guided by the lodestar of foreseeability, the court suggested as 'factors to be taken into account' whether (1) the plaintiff was located near the scene of the accident, (2) the shock resulted from the sensory and contemporaneous observance of the accident, and (3) the plaintiff and the victim were closely related. After a prolonged interval of testing these guide-lines, the court was prepared to abandon even the requirement that the shock resulted in 'physical injury' in the sense of some objective manifestation, like an organic injury such as miscarriage or a heart attack, or at least severe psychiatric trauma. In the *Molien* case, confessing that 'the unqualified requirement of physical injury was no longer justified', it allowed a husband to recover for his anguish at the breakdown of his marriage after his wife had been mistakenly diagnosed as suffering from syphilis.[196]

[193] *Borer* v. *American Airlines, supra*, at 452.
[194] See Prosser and Keeton, *supra*, § 54; Harper, James, and Gray, *supra*, § 18.4.
[195] 68 Cal. 2d 728, 441 P. 2d 912 (1968).
[196] *Molien* v. *Kaiser Foundation Hospitals*, 27 Cal. 3d 916, 616 P. 2d 813 (1980). The majority opinion for a sharply divided court (4:3) was delivered by Mosk, J., who had unsuccessfully advocated a cause of action for loss of

Invoking its now familiar constitutional vocabulary, the court held that the classification was both over- and under-inclusive[197] when viewed in the light of its purported purpose of screening out false claims. It was over-inclusive in permitting recovery on proof of physical injury, however trivial; it was under-inclusive in mechanically denying access to the courts for 'serious mental distress' short of injury. Besides, it encouraged extravagant pleadings and distorted testimony. The new standard had been admitted long ago in cases of *intentional* infliction of mental disturbance[198] and more recently for relational claims by spouses.[199]

The practical result of this decision was to abandon the last judicial control over juries in these cases and to recognize a cause of action for negligent 'pain and suffering' without physical injury.[200]

3.2.1.5 Periodical payments

American law shares with English law the common law tradition of lump sum awards, once and for all, for both past and future losses[201] sustained in a single accident.[202] Prominent among its disadvantages is the necessary guesswork as to future health, life expectancy, inflation, and other vicissitudes. Whereas in England and other Commonwealth jurisdictions this used, until recently, to provide a convenient

---

consortium by parents and children in *Borer* (*supra* n. 187). On another point, the court also abandoned its earlier 'guidelines' that the plaintiff be a 'percipient witness to the injury to the third party' on the ground that here the plaintiff was a 'direct victim'.

[197] *Supra*, ch. 3, at n. 18.

[198] *State Rubbish Collectors Association* v. *Siliznoff*, 38 Cal. 2d 330, 240 P. 2d 282 (1952).

[199] *Supra*, n. 189.

[200] See also *Dartez* v. *Fibreboard Corp.*, 765 F. 2d 456 (5th Cir. 1985), allowing recovery to a former insulation worker for mental anguish from his fear of developing an asbestos-related disease, although he failed to show reasonable medical probability of developing such a disease. *Contra: Payton* v. *Abbott Labs*, 386 Mass. 540, 437 NE 2d 171 (1982).

[201] See Harper, James, and Gray, *supra*, § 25.2.

[202] American law takes a strict stand against 'splitting' causes of action. See *Restatement (Second) of Judgments* § 24 (1982) 'with respect to all or any part of the transaction, or series of connected transactions, out of which the action arose'.

lever for heavily discounting the award, the American jury
has tended to give the plaintiff the benefit of such doubts,
reinforced by the widely prevailing practice of deciding such
issues on an all-or-nothing scale on a balance of probabilities
rather than evaluating the chance and decreasing the award, as
under the English practice, by a corresponding percentage.[203]
The plaintiffs' bar has been less than enthusiastic about
replacing this system by periodical payments, their self-interest
in the matter of fees being quite candidly voiced. It can, and
has been, taken care of, however, by insistence on 'up front'
payment of attorney's fees. Such has become the invariable
practice in so-called 'structured settlements' which provide for
the payment of an annuity, either fixed or variable, often with
a smaller lump sum at the start or at stated intervals.[204] Their
attraction, especially for cases of very severe injury, has been
enhanced by exempting the periodical payments (in contrast
to income from invested lump sums) from income tax.[205] The
reform legislation, repeatedly mentioned, designed to reduce
the cost of medical liability also provides in many states that
an award exceeding a stated sum (like $50,000) in future
damages be converted, at the request of either party, into a
judgment for periodical payments.[206] The Uniform Law
Commissioners' Model Periodical Payment of Judgments Act,
which has not been adopted anywhere yet, respects contractual
stipulations by the attorney that he be paid in a lump sum
even if the client elects to be paid for future damages by
instalments.[207]

[203] See King, 'Causation, Valuation, and Chance in Personal Injury Torts
Involving Preexisting Conditions and Future Consequences', 90 *Yale L. J.*
1353 (1981). For the Commonwealth practice see Cooper-Stephenson and
Saunders, *Personal Injury Damages in Canada*, ch. 3 (1981); Fleming, *Law of
Torts* 205–6 (7th edn. 1987).

[204] See generally *Settlements Including Deferred Payments* (Practicing Law
Institute) (1984).

[205] Periodical Payment Settlement Act of 1982, Pub. L. No. 97–473, 96
Stat. 2605 (1983). See Frolik, 'The Convergence of IRC § 104(2)(a), *Norfolk
and W. Ry.* v. *Liepelt* and Structured Tort Settlements: Tax Policy "Derailed"',
51 *Ford. L. Rev.* 565 (1983).

[206] *Supra*, ch. 3, at n. 36. Similar provisions are contained in some versions
of proposed products liability statutes.

[207] 14 ULA § 7 Comment.

3.2.2 Substantive law

The incentives of the contingent fee are not exhausted by their influence on rules of damages. That their effect on substantive law is less explicit or demonstrable does not detract from the overwhelming consensus that the contingent fee is one of the principal engines driving the judicial activism that has become so characteristic a feature of the modern American tort scene.

We noted earlier the greater incentive for plaintiff's attorneys in the United States to assume the risk of low probability claims; this for the reason that, unlike the individual litigant under the English rule, they are able to spread the cost of failure among their successful clients. Speculative litigation thus becomes feasible and taps a new, if more risky, source of investment. Even if at first the prospect of success is small, the occasional victory furnishes the impetus for continuing to lean on the courts to expand the frontiers of legal liability. In time, this sequence can transform, and has transformed, judicial attitudes regarding the proper role of the judiciary in bringing about legal change. Quite obviously, other factors are also at work to explain the contemporary phenomenon of judicial activism. Some of these have already been noted: the un-responsiveness of other organs of law reform, principally of legislatures; the widespread recruitment of judges from the ranks of the plaintiffs' bar; the pervasive spirit of social welfarism which seems to favour imposing the cost on corporate defendants via tort law rather than on the taxpayer via social security contributions; and so on. All the same, the litigation incentive of the contingent fee has played an undeniably important, more than subsidiary, part in this scenario.

It would be unnecessary here to recapitulate the sequence of decisions that have eliminated the traditional immunities, replaced the defence of contributory negligence by comparative fault, introduced strict liability for defective products, and vastly extended the range of duties of affirmative care and of proximate cause in the search for financially responsible defendants.[208]

[208] *Supra*, chs. 1 and 2.

3.3 Conclusion

One's view of the contingent fee is closely bound up with that of the role of law and lawyers in a changing society. The post-war world has been marked by popular demand for sharply expanded private rights and democratic participation in decision-making processes. Access to justice for the poor and underprivileged is only one item on that social agenda. In contrast to most other countries, including Britain, which see the function of Legal Aid in terms only of a static and minimalist law, in the United States the socially most significant, if controversial, role of legal assistance programmes has been the collective challenge of social abuse and the expansion of rights, including civil rights, of the individual. 'Public interest litigation' has become an accepted avenue for social change.[209] Public support for it is reflected not only in tax support at federal and local level, but in the legion of statutes which authorize award of legal fees to the prevailing party. In fixing such fees the contingency factor has attained recognized status.[210] Traditional tort litigation, under the spur of contingent fees, has long been advancing the same social goals.

Thus the contrast between the United States and other countries involves different perceptions regarding the proper avenues for bringing about legal change. Americans, preferring dispersed political power, look to the courts as much as to legislatures for implementation of changing social purposes; indeed, liberals and radicals have come to look principally to the courts rather than to legislatures for a receptive ear to their programmes. This tendency has put a premium on private law initiatives and on the contingent fee to fuel it. Other countries share neither this constitutional perspective nor nearly the same radical orientation. For them, ordered social change is within the exclusive province of democratically elected legislators expressing a majoritarian will; judges both lack democratic credentials and are by training and association unqualified and averse to becoming agents for social and political change instead of merely guarding the rights of individuals in a conservative tradition.

[209] See Chayes, 'The Role of the Judge in Public Law Litigation', 89 *Harv. L. Rev.* 1281 (1976).   [210] *Supra*, at nn. 14, 43.

7

MASS TORTS

Mass accidents have become a familiar incident of the modern way of life, a by-product of advancing technology in the production, distribution, and use of toxic agents, dangerous pharmaceuticals, fast modern transport, and other hazardous activities. A single type of product like asbestos[1] or the Dalkon Shield,[2] released on a mass market by one or numerous manufacturers, may inflict injury or disease on a vast multitude of consumers or their offspring. Or a single accident, like an aeroplane collision, explosion, or escape of poison gas, may bring injury or death to thousands and dislocation to a whole region. The first is sometimes called a mass products case, the second a mass accident. Both entail injury to multiple victims and present adjudicatory problems very different in magnitude and therefore in kind from those faced in routine accidents. The traditional method of case-by-case adjudication and the applicable principles of substantive law, still largely based on an individualistic philosophy of 'corrective justice' between man and man, are unequal to this challenge.

The congeries of problems is starkly identified by Judge Spencer Williams:

To what extent, if any, must the litigation preferences of each of the thousands of plaintiffs in a 'big case' yield to a more socially optimal, cost-efficient form of representative adjudication? Is our traditional model, pitting one plaintiff's gladiator against one defendant's gladiator, an outmoded and overly expensive means to redress similar injuries inflicted by the same misconduct upon multiple plaintiffs? Must there be dollar-by-dollar liability accumulated by society's producers until the business, or even an entire industry, is forced into bankruptcy or, is it more sensible to have a forum in which all

[1] *Infra*, n. 62.
[2] See Mintz, *At Any Cost: Corporate Greed, Women and the Dalkon Shield* (1985).

potential ramifications upon workers, owners and the future course of product development of the court's decision are aired? Must a single defendant be subjected to numerous and conflicting punitive damage awards, even when such awards threaten, as a legal or practical matter, to deprive future litigants of such recovery, or any recovery at all? Is it fair that a third of any recovery received in early litigation go to the early plaintiffs' attorneys when injured later plaintiffs may be left without practical means of redress? Is it effective or efficient use of juridical resources to subject a judge to the tedious and frustrating task of presiding over identical law suits, or even to distribute these cases throughout the court system to occupy calendars in many courts? Is this one-by-one adjudication the fastest, most equitable way to permit all the injured to recover?[3]

Other common-law countries have not been spared by mass disasters any more than the United States, but for various reasons their courts have not been significantly confronted with the consequential legal problems. In the Thalidomide tragedy, for example, the prevailing doubts in England concerning recovery for pre-natal injuries caused the problem to be shifted to the political forum in order to enlist Parliamentary pressure for settlement.[4] In other incidents, like the Turkish Airline crash outside Paris, claimants went 'forum shopping' and sought American jurisdiction because it offered more promising procedural facilities and surer as well as higher awards than their countries of domicile.[5] In consequence, American courts have taken the brunt of the assault from victims of global disasters and moved further than others in

---

[3] Spencer Williams, 'Mass Tort Class Actions: Going, Going, Gone', 98 FRD 323, 324–5 (1983). The author protested the reversal of his class certification in the *Dalkon Shield* case (*infra*, n. 12).

[4] See Insight Team of the *Sunday Times* of London, *Suffer the Children: The Story of Thalidomide* (1979).

[5] See *In re Paris Air Crash of March 3, 1974*, 399 F. Supp. 732 (C. D. Cal. 1975). On the whole American courts have been hospitable to such forum shoppers, reluctant to invoke *forum non conveniens* and favouring the application of American substantive law against American corporate defendants. The US Supreme Court sought to reverse this trend in *Piper Aircraft Co.* v. *Reyno*, 454 US 235 (1981) (in particular, 'more favourable law' not relevant); followed in *In re Union Carbide Corp. Gas Plant Disaster*, 634 F. Supp. 842 (SDNY 1986) (declining jurisdiction in the Bhopal case). See also *Asahi Metal Industry* v. *Superior Court*, 55 LW 4197 (US Sup. Ct. 1987) (denial of due process to *foreign* manufacturer).

experimenting with ways to cope. This phenomenon itself highlights the fact that the problem facing American courts in these cases is, in significant measure, the harvest of their own ambitious expansion of substantive rights. It is the story of the Sorcerer's Apprentice. Instead of being wary of pushing principles of tort liability beyond the margin of procedural capacity,[6] they have been driven to the more desperate remedy of modifying traditional procedure and rules of proof in order to cope with the resulting flood.

I will report first on procedural innovations, secondly on modifications of rules of proof.

1. PROCEDURE

Two procedural devices for facilitating mass litigation deserve consideration: consolidation and class actions.[7] The following discussion will focus primarily on the Federal Rules of Civil Procedure, partly because they furnished a model for state jurisdictions, partly because most mass tort litigation occurs before federal courts in the exercise of diversity jurisdiction, i.e. suits between citizens of different States or between a citizen and an alien.[8]

---

[6] See, e.g., *State of Louisiana, ex rel. Guste* v. *M/V Test-bank*, 752 F. 2d 1019 (5th Cir. 1985) where Gee, J., reinforced the majority decision against tort recovery for purely economic loss on the ground that 'the dispute resolution systems of courts are poorly equipped to manage disasters of such magnitude and . . . we should be wary of adopting rules of decision which, as would that contended for by the dissent, encourage the drawing of their broader aspects before us' (at 1032).

[7] See generally Transgrud, 'Joinder Alternatives in Mass Tort Litigation', 70 *Cornell L. Rev.* 779 (1985). The simplest and today no longer controversial multi-party procedure is of course ordinary joinder, which may be suitable where the numbers are small.

[8] But not only may there be class actions in State courts, but individual members of a federally certified class may pursue independent actions in State courts, thereby adding another impediment to a comprehensive and uniform disposition of mass claims: see *infra*, n. 31. Prior to *Shutts* (*infra*, n. 38) State class actions of national dimension were foreclosed.

1.1 Consolidation

The most common procedure in mass litigation has been to consolidate actions for pre-trial discovery and sometimes for joint trial of some or all related claims, the only prerequisite being a single common factor of law or fact.[9] Also, since 1968 related cases pending in different districts of the federal judicial system can be transferred to a single district for pre-trial proceedings. The Multidistrict Panel may order such transfer of civil actions involving common questions of fact for the convenience of the parties and witnesses and to promote the just and efficient conduct of pretrial motions and discovery.[10] Besides antitrust and environmental cases, tort actions are peculiarly suitable for MDL (multi-district litigation) treatment. Aircraft disasters[11] and products liability[12] cases have been prominent.

Pre-trial proceedings can vitally affect the outcome of the litigation. Far-ranging discovery, including depositions of the parties and witnesses, tends to flush out the relevant evidence; motions result in rulings on many substantive questions of law, including motions of summary judgment that may terminate the litigation.[13] Although MDL's primary function

[9] Fed. R. Civ. Proc. 42.

[10] 28 USC § 1407(a). See Weigel, 'The Judicial Panel on Multidistrict Litigation, Transferor Courts and Transferee Courts', 78 FRD 575 (1978); Wright, Miller, and Cooper, *Federal Practice and Procedure* § 3861, vol. xv (1976).

[11] e.g., *In re Air Crash Disaster at Boston, Mass. on July 31, 1973*, 399 F. Supp. 1106 (D. Mass. 1975); *In re Paris Air Crash, supra.*

[12] e.g., *In re A. H. Robins Co., Inc. 'Dalkon Shield' IUD Products Liability Litigation*, 406 F. Supp. 540 (JPMDL 1975); *In re 'Agent Orange' Product Liability Litigation*, 506 F. Supp. 762 (EDNY 1980).

[13] Consider this example: 'All of the motions have been copiously documented and excellently briefed and argued. ... The matters submitted and not yet decided concern the constitutionality and application of the Warsaw Convention of 1929, the Hague Protocol, and the Montreal Agreement; the value of gold French francs of 1929 to be considered as against the official value of the gold French frank today, if the Warsaw Convention applies; separate trial of issues and parties; applicability of the California Consumer Credit Act; a motion for summary judgment by defendants against the plaintiffs' claim for punitive damages; and choice of law on liability and damages.' *In re Paris Air Crash, supra*, at 736–7. Qualified under Rule 42 (28 U. S. C. § 1407: 69 FRD 310 (C. D. Cal. 1975). Some of the questions were decided in

is to dispose of pre-trial motions, in practice it often serves to terminate litigation. The persuasive powers of the transferee court judge can be usefully deployed to manœuvre the parties into concessions and eventually into settlement.[14]

Scattered litigation will also be avoided if the actions are eventually consolidated for trial. This has been ordered on a few occasions notwithstanding contrary legislative intent and the wishes of some of the litigators.[15] Most plaintiffs' attorneys, while welcoming the economies of aggregative pre-trial procedure, strongly prefer to conduct their own trial rather than commit trial or settlement to group adjudication.[16] To illustrate: in a group of consolidated Bendectin cases the judge ordered a jury trial of the discrete question whether the morning sickness drug had caused genetic deformities in offspring. When the jury returned a defendant's verdict, not being persuaded by the plaintiffs' statistical evidence, their lawyers protested that the procedure was unfair because it deprived them of the opportunity to show the jury the sort of injuries their clients had suffered and to talk about prior acts of alleged misconduct by the manufacturer.[17]

On the other hand, the cost of subordinating adjudicatory efficiency to the convenience of protesting parties can be substantial. Consolidated trial would preclude opting-out, a choice that is often available even in class actions; it would also favour statistical proof of causation unsuitable for individual litigation. Thus in one well-known instance, the judge's hope of invoking a theory of 'enterprise liability' against all manufacturers of a particular product was eventually dashed by his having to remand various claims to the different districts on grounds of overriding convenience.[18]

622 F. 2d 1315 (punitive damages); 420 F. Supp. 880 (common law wives); 399 F. Supp. 732 (choice of law).   [14] As in *In re Paris Air Crash, supra.*

[15] See Trangsrud, *supra*, at 804-9. These were mostly antitrust actions.

[16] See, e.g., Rheingold, 'The MER/29 Story: An Instance of Successful Mass Disaster Litigation', 56 *Calif. L. Rev.* 116 (1968), where a multidistrict panel had been denied mainly because too many claims were before State courts, but the parties co-operated on a voluntary basis on pre-trial matters.

[17] 'Bendectin Verdict Doesn't End Suit', *Nat. L. J.*, 25 Mar., 1985, at 3, col. 2.

[18] *Chance* v. *I. E. Du Pont*, 371 F. Supp. 439 (EDNY 1974). See *infra*, n. 88.

1.2. Class actions

Class actions are apt to be much more coercive and controversial in the torts context. Their attraction in point of adjudicatory efficiency is bought at the cost of restrictions on individual litigants', particularly plaintiffs', freedom of manœuvre and, as they are apt to claim, their right to due process. Before explaining these conflicting perceptions, it is necessary to outline the statutory conditions for class actions.[19]

Federal Rule 23(a) contains four prerequisites: numerosity, commonality, typicality, adequacy.

(1) The class is so numerous that joinder of all members is impracticable, (2) there are questions of law or fact common to the class, (3) the claims ... of the representative parties are typical of the claims of a class, and (4) the representative parties will fairly and adequately protect the interests of the class.

In addition, the class must qualify in one of three categories, only the first and third of which are relevant to tort claims for damages. The first—a Rule 23(b) (3) classification—applies

if the court finds that questions of law or fact common to the members of the class *predominate* over any questions affecting only individual members, and a class action is *superior* to other available methods for the fair and efficient adjudication of the controversy.[20]

According to a widely held view, which is also shared by Commonwealth courts,[21] mass tort cases do not qualify because claims for damages, unlike injunctions in antitrust cases and

---

[19] See generally Friedenthal, Kane, and Miller, *Civil Procedure*, ch. 16 (1985); Marcus and Sherman, *Complex Litigation*, ch. 4 (1985). Of special interest to English readers is Yeazell, 'Group Litigation and Social Context: Toward a History of the Class Action', 77 *Colum. L. Rev.* 866 (1977), which traces the history of the class action in three distinct roles.

[20] The matters pertinent to the finding include: (1) The interest of members of the class in individually controlling the prosecution or defence of separate actions; (2) the extent and nature of any litigation concerning the controversy already commenced by or against members of a class; (3) the desirability or undesirability of concentrating the litigation of the claims in the particular forum; (4) the difficulties likely to be encountered in the management of a class action.

[21] *Naken* v. *G. M. of Canada*, [1983] 1 SCR 72. See generally, Ontario Law Reform Commission, *Report on Class Actions* (1982).

the like, are predicated upon the unique features of each individual case, affecting causation no less than the *quantum* of damages.[22] Especially products liability, in contrast to mass accident, cases are not based on a single, but a series of, discrete events. The particular circumstances of plaintiff's use of the product and personal knowledge of the danger will almost always differ, as may the governing law and statute of limitations. There may be some common questions, such as whether a particular drug could cause the alleged injuries or whether the manufacturer(s) had or should have had knowledge, but these do not necessarily or unequivocally 'predominate':[23] 'an action nominally conducted as a class action would degenerate in practice into multiple law suits separately tried.'[24] Finally, opposition to class certification can enlist the specific mandate that the court give weight to 'the interest of members of the class in individually controlling the prosecution or defense of separate actions.'[25] Although under this category membership of the class would not be mandatory (as it might be under the second category),[26] allowing for 'opting-out', certification has therefore been mostly denied.[27]

Proponents of the class action deplore this tendency for

[22] Separate adjudication of damages is seen as no obstacle to certification of antitrust and securities class actions.

[23] 'Predominate' involves, of course, a subjective judgment. A representative view is that '[a]lthough common questions need not be dispositive of the entire class action, their resolution should at least provide a definite signal of the beginning of the end': *Mertens* v. *Abbott Laboratories*, 99 FRD 38, 41 (DNH 1983).

[24] Advisory Committee Notes to Proposed Rules of Civil Procedure, 39 FRD 69, 103 (1966). This was the Committee's reason for concluding that 'a "mass accident" resulting in injuries to numerous persons is ordinarily not appropriate for a class action'. That the Note invokes Weinstein's article in 9 *Buff. L. Rev.* 433, 469 is ironic considering his later enthusiasm for class actions in the *Agent Orange* litigation. Also ironically, courts have overcome this concern in mass accident cases, which the Committee addressed, unlike in products cases, which it did not.

[25] Rule 23(b)(3)(A). See *supra*, n. 20.

[26] *Infra*, n. 38.

[27] *In re Northern District of California 'Dalkon Shield' IUD Product Liability Litigation*, 693 F. 2d 847 (9th Cir. 1982); *Payton* v. *Abbott Labs*, 100 FRD 336 (D. Mass. 1983) (DES); *Yandle* v. *PPG Industries, Inc.*, 65 FRD 566 (E. D. Texas 1974) (employees of asbestos plant); *Mertens* v. *Abbott Laboratories*, *supra* (DES).

giving insufficient attention to the possibility of certifying a class on *any* common issues: even if individual issues outnumber common ones, that should not preclude an 'issues only' class action.[28] The sheer avalanche of claims in recent mass litigation has had a noticeable effect on judicial attitudes. After prolonged discouragement by appellate courts, the need to sacrifice traditional procedures for the sake of avoiding a complete breakdown of the judicial system has become more widely recognized, even among plaintiffs' attorneys who have most to lose except the opportunity to get to trial. In the *Agent Orange*[29] and *Abestos*[30] litigation, class actions were accordingly certified with appellate approval, on specific issues such as the defectiveness of a generic product and the state-of-the-art defence. Mass disaster cases, like the *Skywalk* case involving the collapse of mezzanine walkways in a hotel, are even more suitable for class management than products cases because of the typical identity of the causation issue. In the leading case, after mandatory certification was thwarted by an appellate court, voluntary certification eventually led to a successful settlement.[31]

A second category for potential application to mass torts— Rule 23(b)(1)—qualifies class actions in order to avoid a scramble for a limited fund. It contemplates a situation in

[28] See Spencer Williams, *supra.*

[29] *In re 'Agent Orange' Product Liability Litigation*, 100 FRD 718 (EDNY 1983), aff'd 725 F. 2d 858 (2nd Cir. 1984). Agent Orange, a chemical containing dioxin, was used by the US Army in Vietnam to defoliate the countryside and has been accused of exposing millions to toxic injury. See Schuck, *Agent Orange On Trial* (1986).

[30] *Jenkins* v. *Raymark Industries, Inc.*, 782 F. 2d 469 (5th Cir. 1986) (limited to state-of-the-art defence) (5,000 cases were pending in the 5th Circuit alone); *In re School Asbestos Litigation*, 789 F. 2d 996 (3rd Cir. 1986) (with a potential of 140,000 claims).

[31] In *In re Federal Skywalk Cases*, 680 F. 2d 1175 (8th Cir. 1982), involving a single accident, *mandatory* certification originally granted (93 FRD 415 (W. D. Mo. 1982)) was reversed on appeal for violating the Anti-Injunction Act by interfering with claimants in State courts. This reversal did not preclude the later voluntary certification both in federal and state courts. The appeal decision has been severely criticized for in effect throwing the burden on state courts: Wright and Colussi, 'The Successful Use of the Class Action Device in the Management of the *Skywalks* Mass Tort Litigation', 52 *U. Miss.-Kansas City L. Rev.* 141 (1984). This view is contested by Morris and See, 'The Hyatt Skywalks Litigation: The Plaintiffs' Perspective', ibid., at 246.

which there are multiple claimants to a limited fund and a risk that if litigants were allowed to proceed on an individual basis, those who collect first will deplete the fund and leave nothing for latecomers. For example, liability for a nuclear accident is limited at present to $500m. under federal legislation.[32] More generally, in tort actions, the defendant's assets might constitute such a limited fund, especially in face of punitive damages.[33] Such claims by thousands of litigants, if successful, could quickly exhaust even a strong corporate defendant's assets or courts might hold at some point that the defendant has been punished enough.[34] In either case latecomers would get short shrift.

All the same, certification of classes for the purpose of punitive damages has not met with much success. It is not seriously disputed that Rule 23(b)(1) may be applied to mass tort cases. But hitherto appellate courts have mostly reversed certification for one or other flaw.[35] One court also postulated the severe test that separate punitive damage claims must 'inescapably' affect later claims.[36] But this seems incompatible with the statutory requirement that there merely be a 'risk'; at any rate Judge Weinstein in the *Agent Orange* case contented himself with the lesser standard of 'substantial probability—that is, less than a preponderance but more than a possibility'.[37]

These limited fund cases pose an additional problem in that

[32] Price–Anderson Act, 42 USC § 2210. See Note, 'Nuclear Power and the Price–Anderson Act: Promotion Over Public Protection', 30 *Stan. L. Rev.* 393 (1977).

[33] See Note, 'Class Actions for Punitive Damages', 81 *Mich. L. Rev.* 1787 (1983).

[34] This, the 'overkill' argument was pressed in *In re School Asbestos Litigation*, *supra*, but was deflated on several grounds, among them that it was not accepted by many courts.

[35] *In re Dalkon Shield, supra*, n. 27 (no plaintiff or defendant supported class certification); *In re Federal Skywalk Cases, supra* (mandatory certification violated Anti-Injunction Act); *In re Bendectin Products Liability Litigation*, 749 F. 2d 300 (6th Cir. 1984) (no findings as to limited fund). *In re School Asbestos Litigation, supra* (class was 'underinclusive'; thus the action would not have silenced non-school claims). If the view prevails that Rule 23 is not an exception to the Anti-Injunction Act, its benefits can be largely undercut: see Wright, Miller, and Kane, *Fed. Practice and Procedure*, vol. 78 § 1798.1.

[36] *In re 'Dalkon Shield', supra*, n. 27, at 852.

[37] *In re 'Agent Orange', supra*, n. 29, at 726.

certification, to be effective, must be mandatory, precluding members from 'opting out'. Absent members need not even be notified. This raises the question whether jurisdiction can be entertained over absent class members who lack substantial contact with the state,[38] as well as the prohibition against federal courts interfering with proceedings in State courts.[39] Moreover, how can the case be definitively closed without providing for future claimants, especially troublesome in case of latent diseases?

1.2.1 Evaluation

Class actions for mass torts have attracted strong advocates as well as strong opponents. This partisanship is not always identical with the division between defendants and plaintiffs. Although defendants have usually much to gain from a single adjudication, some may perceive tactical advantages by taking on claimants separately, in the hope of negotiating cheaper settlements or even discouraging claims altogether. Plaintiffs' attorneys are usually in the forefront of opposition, but their interests are not always identical. Those with large claims generally hope to do better by going it alone, while the smaller fry may expect to gain from the greater bargaining power of the class.

1.2.1.1   Efficiency: issue preclusion

Adjudicatory efficiency is the central claim for class actions, procedural fairness against it. Separate litigation of numerous similar claims entails enormous waste of resources, whereas findings on common questions of law or fact bind all members

---

[38] At least in non-mandatory actions, such contacts are not required by the 14th Amendment (due process) for plaintiffs because, unlike defendants, they are not exposed to an undue burden: *Phillips Petroleum Co.* v. *Shutts*, 472 US 797 (1985). See Miller and Crump, 'Jurisdiction and Choice of Law in Multistate Class Actions After *Phillips Petroleum Co.* v. *Shutts*', 96 *Yale L. J.* 1 (1986); Seltzer, 'Punitive Damages in Mass Tort Litigation: Addressing the Problems of Fairness, Efficiency and Control', 52 *Ford. L. Rev.* 37 (1983); Byer, 'National Mandatory Class Actions: Key Questions Remain Unanswered', *Nat. L. J.*, 30 Sept. 1985, p. 19.

[39] *Supra*, n. 31.

of a class, thereby precluding endless relitigation.[40] Even so-called 'test actions', which are widely touted as an adequate alternative to class actions,[41] have no *res judicata* effect and at best only persuasive force on non-parties. Admittedly, American courts have gone much further than the English in extending the scope of issue estoppel. The traditional view, still adhered to by English courts,[42] is that only parties and their privies are bound by, or can take advantage of, a prior judgment. While the first proposition, that non-parties cannot be bound, is if anything reinforced by the US constitutional requirement of due process,[43] the principle of mutuality underlying the second proposition, that only privies can take advantage of a judgment, no longer commands unqualified support. By giving prior parties an opportunity to relitigate in the hope of a different outcome, mutuality sacrifices judicial economy and raises the possibility of inconsistent results.[44] Not surprisingly, symmetry has not been sufficient to defend the rule in its entirety against critics.

Accordingly, varying modifications of the principle of mutuality have been allowed, the most common being where the party against whom estoppel is being asserted was on the offensive in the first action and the party asserting it is on the defensive in the second.[45] There is little sympathy for a losing claimant, who has had his day in court, seeking to relitigate the same issues offensively against other adversaries. On the

---

[40] *Hansberry* v. *Lee*, 311 US 32 (1940). See Friedenthal, Kane, and Miller, *supra*, § 16.8.

[41] See *Report on Class Actions*, *supra*, at 86. A recent illustration of a test case for 20,000 industrial victims is *Thompson* v. *Smiths Shiprepairers* [1984] Q.B. 405

[42] See Cross, *Evidence*, ch. 13 (5th edn. 1979).

[43] e.g. in *Hardy* v. *Johns–Manville Sales Corp.*, 681 F. 2d 334 (5th Cir. 1982) asbestos defendants were held not estopped by findings against different defendants in previous litigation.

[44] See Friedenthal, Kane, and Miller, *supra*, § 14.14; *Restatement, Judgments, Second* § 29 (1982).

[45] This was the situation in the seminal case of *Bernhard* v. *Bank of America*, 19 Cal. 2d 807, 122 P. 2d 892 (1942) (Traynor, J.). So also in *In re 'Dalkon Shield' Punitive Damages Litigation*, 613 F. Supp. 1112 (E. D. Va. 1985), a prior denial to certify a class action at the behest of the defendant (*supra*, n. 27) precluded the same defendant from raising the issue once more as against different plaintiffs who were using preclusion defensively.

other hand, many courts have refused to allow non-parties to
benefit from issue preclusion offensively. This distinction rests
on two reasons: first, to abandon mutuality in the latter
situation would encourage plaintiffs to sit out the first action:
if it turned out successful, they would take advantage of it;
but if unsuccessful, they would repudiate it on due process
grounds. Secondly, in a mass accident, if the first judgment
went against the defendant, all claimants would take advantage
of it, but if it went for the defendant, they could relitigate.[46]
Under a jury system, inconsistent verdicts are not infrequent,
with the result that claimants would pursue separate actions
until one verdict ended in their favour. Indeed, if we leave
issues like defectiveness of a product design to the jury at all—
the wisdom of which has been disputed[47]—it must be on the
assumption that reasonable minds can differ on such a
judgmental question, and it would be self-defeating to defer
to any single verdict as definitive and pre-emptive.[48] But does
this not weigh as heavily against a one-shot decision in the
framework of a class action? One difference is that in a class
action all parties represented are bound by a determination
either way, so that fairness and due process are not impaired
in the same measure by issue preclusion. Another is that the

---

[46] See *Parklane Hosiery* v. *Shore*, 439 US 322, 330 (1978). It was on this
among other grounds that *Hardy* v. *Johns–Manville Sales Corp.*, *supra*, decided
against issue preclusion in asbestos litigation. See Currie, 'Mutuality of Collateral Estoppel: Limits to the Bernhard Doctrine', 9 *Stan. L. Rev.* 281 (1957);
Flanagan, 'Offensive Collateral Estoppel: Inefficiency and Foolish Consistency', 1982 *Ariz. St. L. J.* 45.

[47] *Supra*, ch. 2, at n. 138.

[48] *Restatement, Judgments, Second* § 29 which follows the more fluid New York
test, states in comment g: 'The circumstances attending the determination of
an issue in the first action may indicate that it could reasonably have been
resolved otherwise ... Resolution of the issue in question may have entailed
reference to such matters as the intention, knowledge, or comparative responsibility of the parties in relation to each other ... In these and similar
situations, taking the prior determination at face value for purposes of the
second action would extend the effect of imperfections in the adjudicative
process beyond the limits of the first adjudication....'

But in *Kaufman* v. *Eli Lilly & Co.*, 65 NY 2d 449, 482 NE 2d 63 (1985) the
NY court invoked issue estoppel against the same defendant on behalf of a
different DES claimant when the first-tried case was 'roughly typical' of the
other. One contrary jury finding in another jurisdiction against a different
manufacturer of the same drug was not sufficient to deny preclusive effect.

risk of aberrational outcomes is reduced because a class action stimulates maximum effort by plaintiffs and the judge acts as protector for absent class members.

1.2.1.2   Fairness and ethics

None the less, procedural fairness stands at the top of the objections against class actions, usually voiced by plaintiffs. Their contention is that each tort plaintiff should have the right to prosecute his own action and be represented by an attorney of his choice. Class actions were intended to assist the prosecution of 'individually nonrecoverable claims',[49] i.e. claims which would not justify the expense to an individual of independent litigation because the plaintiff's own stake was too small or non-provable, as in many environmental and consumer protection cases. Such, however, is not true of tort claims where contingent fees furnish the key to the courthouse. To opponents this plea sounds specious and hypocritical. In class actions a management committee and 'lead counsel' usually represent the class, probably with greater skill and resources than would otherwise be available to most class members.[50] Behind the rhetoric is not the client's, but his attorney's concern about being short-changed on his fees. Usually the court will set the fees with reference to the particular attorney's contribution in time, skill, and risk in relation to the entire enterprise, to the disadvantage of many attorneys whose role may have been little more than that of referral.[51] Indeed, one of the most disenchanting aspects of class actions is the internecine manœuvrings by attorneys for a place in the sun, oblivious of a serious conflict of interest between themselves and their clients.[52]

[49] Note, 'Developments in the Law: Class Actions', 89 *Harv. L. Rev.* 1318, 1356 (1976), distinguishing between the aggregation in class actions of non-viable, individually non-recoverable and individually recoverable claims.

[50] One of the 4 prerequisites of a class action under Rule 23(a) is that 'the representative parties will fairly and adequately protect the interests of the class': *supra*, at nn. 19–20.

[51] e.g. *In re 'Agent Orange'*, 611 F. Supp. 1223 (1985); Riley, '"Agent Orange" Fees Sharply Curtailed', *Nat. L. J.*, 21 Jan. 1985, at 3, col. 2.

[52] A particularly shocking example was the jockeying among the *Bhopal* attorneys: see e.g. Meier, 'Lawyers for Victims of Bhopal Gas Leak Fighting One Another', *Wall St. J.*, 1 May 1986, p. 1.

Ethical concerns posed by contingent fees in general[53] are if anything magnified by class actions. One is the potential conflict of interest between lawyer and client in relation to settlements. In view of the judicial fee-setting formula just referred to, which puts major emphasis on hours of work, the lawyer has usually little to gain and much to lose by going to trial, especially as the court will ordinarily allow nothing if the action is unsuccessful.[54] Most serious of all is the problem of solicitation, which received a great deal of adverse publicity after the Bhopal disaster in 1985. Victims of a mass disaster are easy to locate, there are many of them, and the more clients the lawyer garners, the greater will be his likely financial return. Critics of class actions have often protested that most of them are instigated by attorneys on behalf of clients who individually have little to gain from the litigation.[55] This is especially the case with 'individually nonrecoverable claims' for which class actions were primarily designed, such as consumer fraud cases; it is less true of tort claims which are usually for substantial amounts. These critics are in fact turning the efficiency argument against defenders of class actions, contending that the burden on courts in overseeing such complex litigation, thus unnecessarily stirred up, vastly exceeds imagined economies.[56]

1.2.1.3 Social Activism

However distasteful some manifestations of aggressive solicitation, it must be recognized that the ancient common-law condemnation of maintenance and champerty has long given way in the United States to tolerance; indeed in many instances, as we have seen, the public interest actively encourages private litigation as the most effective form of law enforcement. It does so through generously 'inferring' private

[53] *Supra*, ch. 6, at n. 64.
[54] *Hensley* v. *Eckerhart*, 461 US 424 (1983); *Blum* v. *Stenson*, 465 US 886 (1984).
[55] e.g. *Kline* v. *Coldwell Banker & Co.*, 508 F. 2d 226, 238 (9th Cir. 1974) (Duniway, J., concurring).
[56] This division of opinion cuts across the usual party lines. For example, avowed conservatives have split on whether to encourage (Spencer Williams, *supra*) or discourage (Duniway, J., *supra*) class actions.

causes of action from regulatory legislation,[57] by the numerous statutes permitting award of attorney fees to prevailing plaintiffs in public interest litigation,[58] and, quite evidently, in endorsing the very device of class actions. The larger issue is therefore whether the extra burden, if any, placed on the judicial system by class actions is not a tolerable price to pay for promoting other policy goals. Reducing the volume of litigation to a minimum cannot be a justifiable end in itself; in a democratic society legitimate claims, particularly against powerful interests, should not be barred from access to justice. The noticeable opposition to class actions by many appellate judges seems to be largely motivated by their concern that scarce judicial resources are overtaxed by complex litigation in the pursuit of trivial claims. Yet there is widespread public support for comprehensive enforcement of regulatory legislation (in environmental, security regulation, consumer protection cases) which must be given weight against the burden of class actions on the judicial system.[59] In any event, personal injury claims are substantial by any definition and their suitability for class actions should be considered on their own terms of procedural efficiency *vis-à-vis* individualized proceedings.

The enthusiasm of social activists for class actions is shared by some trial judges whose belief in their comparative efficiency may be coloured by the opportunity they offer for a display of judicial grandeur and a reputation for innovative ideas. An even more radical role for class actions is envisaged by Professor David Rosenberg.[60] Admitting the cogency of the widespread criticism of the tort system as too cumbersome, costly,

---

[57] See *Cort* v. *Ash*, 422 US 66 (1975); *J. I. Case Co.* v. *Borak*, 377 US 426 (1964). Also, e.g., Sunstein, 'Section 1983 and the Private Enforcement of Federal Law', 49 *U. Chi. L. Rev.* 394 (1982).

[58] *Supra*, ch. 6, at n. 13.

[59] See Dam, 'Class Actions: Efficiency, Compensation, Deterrence, and Conflict of Interest', 4 *J. Leg. Stud.* 47 (1975).

[60] Rosenberg, 'The Causal Connection in Mass Exposure Cases: A "Public Law" Vision of the Tort System', 97 *Harv. L. Rev.* 849 (1984). Taking sharp issue with this advocacy of 'privatization of public risks' is Huber, 'Safety and the Second Best: The Hazards of Public Risk Management in the Courts', 85 *Colum. L. Rev.* 277 (1985), who argues that it is in the long run counterproductive to safety to discourage the benefactions of modern technology.

and haphazard to accomplish its accident prevention and compensation objectives[61] with respect to sporadic accidents like automobile collisions, he contends that mass accidents are peculiarly amenable to tort treatment in class actions. Inasmuch as these accidents are frequently the result of deliberate business decisions balancing safety against profits, they should be especially amenable to control through threats of liability; while their massive scale and statistical pre-dictability render them peculiarly suitable for aggregative rather than case-by-case adjudication. This 'public law' ap-proach may justify modifications of the traditional 'private law' approach both in relation to proof and remedies—a theme which we will take up shortly.

1.3 Corporate reorganization

An entirely novel procedure for disposing of mass claims was inaugurated when the giant Johns–Manville Corporation sought protection under Chapter 11 of the Bankruptcy Act against the claims of more than 10,000 personal injury claims by asbestos victims.[62] Chapter 11 contemplates reorganization, rather than dissolution, as a method of helping financially distressed corporations return to viability. What was novel was that this procedure was enlisted for relief not against commercial, but against tort, creditors. A prominent goal of American bankruptcy law is not only to insure equality among (unsecured) creditors, but to rehabilitate debtors. A recent reform increased this tendency by making most tort claims dischargeable, thereby opening the possibility of allowing a corporation to discharge mass tort claims through a Chapter 11 reorganization.[63] Johns–Manville's bold manœuvre im-

---

[61] See e.g. Sugarman, 'Doing Away with Tort Law', 73 *Calif. L. Rev.* 558 (1985); Fleming, 'Is there a Future for Tort?', 44 *La. L. Rev.* 1193 (1984).

[62] See Hensler, Felstiner, Selvin, and Ebener, *Asbestos in the Courts: The Challenge of Mass Toxic Torts* (1985). The dimensions of prior asbestos litiga-tion are indicated in the Appendix to *Jackson* v. *Johns–Manville Sales Corp.*, 750 F. 2d 1314, 1335–41 (5th Cir. 1985). See also from a pro-plaintiff view Brodeur, *Outrageous Misconduct: The Asbestos Industry on Trial* (1985).

[63] See Hillson, 'The Genesis of a New Trend: Chapter 11, Avoiding or Managing Future Liability in Mass Tort Actions', 15 *Capital U. L. Rev.* 243 (1986); Roe, 'Bankruptcy and Mass Torts', 84 *Colum. L. Rev.* 846 (1984);

mediately blocked pending claims, new suits, and execution on judgments, while the corporation continued in business.

The reorganization plan, filed in July 1985, provides for the creation of a permanent fund for the personal injury claimants, consisting of cash and insurance proceeds of $850m., a $1.65b., bond (which will provide $75m. annually) and 50 per cent of the company's common stock, with excess of up to 80 per cent.[64] Recognized are claims of both present and future putative victims.

All interested parties, creditors as well as stockholders, benefit from this arrangement. While some tort claimants will receive less than an award they might have received at the hands of a jury in individual trials, in a case like the present where the claims exceed the defendant's assets[65] this procedure assures a prompter and more equitable distribution than would a free-for-all race for trial. Moreover, the limited assets available for distribution will not be drained by the high cost of traditional adjudication or the haphazard award of punitive damages. Indeed, continued operation of the enterprise is likely to earn profits which can be earmarked for the benefit of victims, thereby increasing their likely share and providing for future putative claimants. Finally, reorganization prevents the total collapse of important enterprises providing essential products and services. Modern technology has immensely increased the risk of mass disasters; and while stockholders cannot claim immunity from responsibility which the law exacts, it would be demanding a gratuitous sacrifice, harmful to the public interest, if an otherwise beneficial enterprise were denied the opportunity of continuing to function.

Reorganization under Chapter 11 would not be justified, however, unless the claims exceeded the resources of the enterprise. Thus most asbestos producers have had to resort to other defensive strategies.[66] Reorganization does not furnish an overall remedy for mass claims.

Note, 'The Manville Bankruptcy: Treating Mass Tort Claims in Chapter 11 Proceedings', 96 *Harv. L. Rev.* 1121 (1983).

[64] See 'The Manville Settlement', *Nat. L. J.* 19 Aug. 1985, p. 30, col. 1. This still left property claims for $86bn..   [65] Estimated at $2bn.

[66] A few other producers followed the Johns-Manville example, as did the A. H. Robins Company, manufacturer of the Dalkon Shield. Many of the

2. BURDEN OF PROOF

One of the most frequent difficulties encountered in mass accident cases relates to proof of causality. It has two aspects: first, the problem of the indeterminate defendant. It is often unclear which of several manufacturers of, say, a drug produced the particular unit of the product that harmed the plaintiff. The generic character of the product, the inconspicuousness of the exposure event, and the long latency period frequently prevent precise identification of the responsible manufacturer. Secondly, the problem of the indeterminate plaintiff. Especially in pollution cases, the plaintiff can often rely only on general statistical information to suggest that the defendant's emission merely increased the number of sufferers beyond those who would have contracted the disease in any event from other human agents or perhaps legally non-responsible background risks. Does this sufficiently identify the plaintiff as one injured, rather than merely threatened, by the defendant?

The traditional requirement that the plaintiff prove causality against each defendant on a balance of probabilities reflects our notions of procedural fairness in the individualized confrontation typical of random accidents. It is argued, however, that this 'rule is neither a rational nor a just means of resolving the systematic causal indeterminacy presented by mass exposure cases.'[67] This postulate calls essentially for modification of conventional substantive law in order to exploit the procedural advantages of class actions in mass tort cases. To what extent has substantive law already bent to this challenge?

---

other producers accepted the 'Wellington Facility', an agreement between manufacturers, their insurers, and plaintiffs' attorneys for an orderly system of compensation (in accordance with tort principles). It is voluntary for plaintiffs but has the advantage of speedier compensation than litigation. See Note, 'The Asbestos Claims Facility: An Alternative to Litigation', 24 *Duq. L. Rev.* 833 (1986).

[67] Rosenberg, *supra*, at 858. Systemic?

2.1 The indeterminate defendant

Modifications of the conventional rule, which places the burden of proof on the plaintiff to identify which one among a group of potential culprits was responsible for his injury, have actually preceded the advent of class actions. Most of these emanated from California, and especially the more radical of them have not, at least not yet, been widely followed elsewhere. While the earlier cases involved random accidents, such as a hunting accident and a mishap during surgery, the problem is destined for a more prominent role in products liability claims involving design defects. The danger element may lurk either in a generic product or in a generic component of a product manufactured by more than one producer. The dramatic extension of strict liability for design defects, previously noted,[68] has greatly increased the opportunities for testing its application to the problem of the indeterminate defendant. This is reflected in the experience of recent mass products cases, like the DES, *Agent Orange*, and asbestos cases.

2.1.1 Alternative liability

The earliest, so-called 'alternative liability' theory originated in the case of *Summers* v. *Tice*[69] where two hunters, using shotguns, fired simultaneously in the direction of the plaintiff, one shot putting out his eye. The court reversed the conventional burden of proof, holding that where a single injury has been inflicted by one or the other of two negligent defendants, but the plaintiff cannot prove which one, it was for each of them to exculpate himself by establishing on a balance of probabilities that he was not the one. The rationale of this decision was that the equities between an innocent plaintiff and two negligent defendants, each one of whom could have caused his injury, favour placing the risk of proof uncertainty on the latter.[70]

---

[68] *Supra*, ch. 2, at 130.

[69] 33 Cal. 2d 80, 199 P. 2d 1 (1948).

[70] *Restatement, Torts, Second* § 433B (1965). The principle has been repeatedly applied in chain collision and water pollution cases, like *Maddux* v. *Donaldson*, 362 Mich. 425, 108 NW 2d 33 (1961); *Landers* v. *East Texas Salt Water Disposal Co.*, 151 Tex. 251, 248 SW 2d 731 (1952). It is also behind

It has been questioned whether this principle should be confined to two defendants, in which case the odds on either one's being the culprit are 50 : 50. Contribution could ensure that each bore 50 per cent of the loss, so that the extent of each one's liability would in effect reflect the probability of his having caused the injury. While we are in general reluctant to accept statistical proof of culpability, particularly on the question of identification,[71] those concerns have much less weight in application to defendants whose negligence has once been established. Moreover, matching the extent of liability to the degree of probable causation is an accepted rule for assessing damages for future contingencies.[72] Thus the chance of future arthritis or epilepsy, even if less than 'more probable than not' (51 per cent?), justifies an award, not for 100 per cent, but for the discounted value of its probability (which may be more or less than 50 per cent). Applying the same rationale to proof uncertainty on causation—both with respect to the question of whether it would have made a difference had the defendant been careful[73] and to the present question of which one of several negligent actors caused the injury—is therefore not as great a departure from conventional premises as might first have appeared. Proportional liability would also dispel concern if the rule in *Summers* v. *Tice* were extended beyond two defendants. For though the chance of any one being the responsible hunter would in that event decrease below 50:50, each one's share of liability would also be reduced proportionately, thus not affecting the equity of the formula.

The preceding argument assumes, however, that all possible culprits are before the court and financially responsible, for

cases which shift the burden of proof to a negligent tortfeasor to show for how much of the damage he is not responsible, but these assume that he has caused some at least of the damage.

[71] See Tribe, 'Trial by Mathematics: Precision and Ritual in the Legal Process', 85 *Harv. L. Rev.* 1329 (1971).

[72] This is well established in English law, not so well in American law. See Cooper-Stephenson and Saunders, *Personal Injury Damages*, ch. 3 (1981); King, 'Causation, Valuation and Chance', 90 *Yale L. J.* 1353 (1981).

[73] See *Herskovits* v. *Group Health Cooperative*, 99 Wash. 2d 609, 664 P. 2d 474 (1983); *Hotson* v. *E. Berks. H. A.* [1987] 2 All E. R. 909 (would the patient's life have been saved by a more prompt diagnosis?). Also King, *supra*.

otherwise proportional distribution might be impaired. It was
for this reason that the California Court declined to apply the
principle against the five DES manufacturers sued in the case
of *Sindell* v. *Abbott Laboratories* where more than 200 had in
fact marketed the drug at the relevant time.[74]

It is not a necessary condition, any more than for the
procedural rule of *res ipsa loquitur*, that the defendants are in
a better position than the plaintiff to know what actually
happened. On occasion, the desire to loosen the defendant's
tongue has played a role in imposing the burden of exculpation
on him. Best known is the famous decision of *Ybarra* v.
Spangard[75] where, in order to break the 'conspiracy of silence',
the burden of proof was shifted to all members of a surgical
team to explain what accounted for an external injury to the
patient's shoulder during an appendectomy. Since none could
explain, all were held liable.[76] The converse, however, does
not apply. Thus in the case of the hunters, neither one had
any more knowledge than the victim as to whose shot had
caused the injury. So also in DES litigation, the fact that the
manufacturers were unable to identify which, among them,
had produced the drugs taken by the plaintiff's mother (indeed
that, if anything, the latter would have been in a better
position to remember) did not by itself preclude the application
of this principle.[77]

2.1.2   Concerted action

The classical theory of joint liability is that of persons acting
in concert, who are held 'jointly and severally' liable for each
other's acts performed pursuant to the agreement to commit

---

[74] 26 Cal. 3d 588, 602–3; 607 P. 2d 924, 930–1 (1980). Also courts rejecting
the *Sindell* market-share solution usually emphasized as its most objectionable
feature that the real culprit might not be among those held liable: see *In re
'Agent Orange' Product Liability Litigation*, 597 F. Supp. 740, 826 (EDNY 1984).

[75] 25 Cal. 2d 486, 154 P. 2d 687 (1944). The Court invoked the analogy
of *res ipsa loquitur* for reversing the burden of proof. Its application in *Anderson*
v. *Somberg*, 67 NJ 291, 338 A. 2d 1 (1975) was much more radical because
there was no comparable link between the surgeon, retailer, and the manufacturer of the surgical tool that fractured.

[76] 93 Cal. App. 2d 43, 208 P. 2d 445 (1949).

[77] *Sindell* v. *Abbott Laboratories, supra*, at 600–2, 607 P. 2d 929–30.

a tort against the plaintiff. The latter is therefore absolved from identifying who of the conspirators actually caused his injury. Although the stock situation contemplates an agreement to injure the plaintiff intentionally, its application to other tortious conduct, while rare, is not in serious doubt. Thus an agreement to engage in conduct fraught with unreasonable risk to the plaintiff will be sufficient, as would perhaps concerted action fraught with strict liability, like marketing dangerous drugs or other products.

Under what conditions this theory could be invoked against manufacturers of a generic drug has been considered with varying results in DES litigation. Following the *Sindell* decision in California (below),[78] all but one court declined to find a common plan or design merely in parallel or imitative conduct. True, express agreement is not necessary and all that is required is a tacit understanding. But mere reliance on each other's testing and promotion methods does not qualify, since that is a common practice in industry and would render virtually any manufacturer liable for the defective products of an entire industry even if he could demonstrate that the product which caused the injury was not made by himself.[79] Even courts prepared to accept 'market-share' responsibility have, as we shall see, consistently dismissed claims against defendants who could not have supplied the injury-causing product because they had not marketed it in that geographical area.[80] Nor have they allowed a claim to proceed on such a theory until the plaintiff had failed to identify the defendant as the source of the particular product.[81] Neither conclusion is compatible with the theory of joint liability for concerted action; indeed both are difficult enough to reconcile with the modified market-share theory.

In only two instances has the theory of concerted action made better headway. In one case the highest New York

[78] *Supra*, n. 74.

[79] Ibid., at 605, 607 P. 2d 932–3.

[80] e.g. *Sindell, supra*, at 612, 607 P. 2d at 937; *Hall, infra*, at n. 85. *Powell* v. *Standard Brands Paint Co.*, 166 Cal. App. 3d 357, 212 Cal. Rptr. 395 (1985) rejected a claim against another manufacturer who had also failed to attach warning to his product, the source of the culpable product being identified.

[81] e.g. *Thompson* v. *Johns–Manville Sales Corp.*, 714 F. 2d 581 (5th Cir. 1983).

Court, without actually deciding whether the theory was appropriate to DES litigation,[82] held that an implied agreement or co-operation could be inferred from consciously parallel conduct, such as common failure to test the drug on pregnant mice.[83] The jury was entitled to find that the one manufacturer who alone was sued had thereby substantially aided or encouraged other DES manufacturers to follow his example. Eight manufacturers had originally filed applications to market DES for pregnancy problems and each had relied on the same research studies. This court apparently did not share the view of the California Court that since the licensing and marketing of drugs was closely regulated by the Food and Drug Administration, no adverse inference could fairly be drawn against applicants who merely followed the procedures stipulated by the FDA.[84]

A forerunner of the concert-of-action theory applied against multiple manufacturers was the ambiguous decision in *Hall* v. *E. I. Du Pont de Nemours*.[85] In one of two consolidated cases thirteen children had been injured in separate incidents by blasting caps. Being unable to identify the particular manufacturer, the plaintiffs sued six domestic producers, comprising virtually the whole industry, with the complaint that they had all failed to attach a suitable warning, that they knew about the risk but jointly decided against labelling and lobbied against it. The innovative federal Judge Weinstein held that these allegations stated a cause of action for joint liability based on joint control of the risk. It was sufficient if the defendants, though acting independently, had adhered to an industry-wide practice with regard to the safety features of the blasting caps. Despite this liberal opening to parallelism, the court hedged in two respects which departed from the traditional theory of joint tortfeasorship. First, by allowing exculpation to any defendant who could prove that he did not manufacture the cap that injured the plaintiff.[86] Secondly,

[82] Consequently, no estoppel on that issue could be invoked against the defendant by other plaintiffs: *Kaufman* v. *Eli Lilly & Co.*, *supra*.
[83] *Bichler* v. *Eli Lilly & Co.*, 55 NY 2d 571, 436 NE 2d 182 (1982).
[84] *Sindell, supra*, at 609-10, 607 P. 2d at 935.
[85] 345 F. Supp. 353 (EDNY 1972).
[86] In the other action consolidated in *Hall*, the court refused to allow joinder of manufacturers whose caps had not injured the plaintiffs: at 382-4.

by cautioning against application to a large number of producers.[87] Moreover, the judge later unravelled his own decision by remanding the individual cases to different districts where the accident had occurred.[88]

In DES litigation, the *Hall* principle of enterprise liability has been consistently rejected on the ground that the fact that hundreds of drug companies entered and left the market over many years fatally weakened the assumption that the defendants jointly controlled the risk.[89] In the *Agent Orange* case, where only seven manufacturers were involved, the same judge who had decided *Hall* declined to find parallelism merely in a deliberate failure by all defendants to reduce the dioxin level of the herbicide used by the US forces in Vietnam. It showed only that the defendants had not taken adequate action, not an industry-wide decision to take inadequate action.[90] On the other hand, parallelism would apply to an understanding among the manufacturers to keep the government in the dark about the danger of dioxin.[91]

2.1.3 Market share

The most innovative theory was launched by the California Court in *Sindell* v. *Abbott Laboratories*.[92] Having rejected all the precedents as unsuitable for application against the more than 300 manufacturers of DES because they would have exposed each of them to joint and several liability for every injury caused by a 'defective'[93] generic drug, the Court

[87] Ibid., at 378. It was for this reason that the *Sindell* court declined to apply *Hall* to the 300 DES manufacturers (*Sindell, supra*, at 609, 607 P. 2d at 935).

[88] *Chance* v. *E. I. Du Pont de Nemours, supra*. One reason was that there was no federal common law of liability (at 448). This was reaffirmed by the Second Circuit Court of Appeals in *In re 'Agent Orange'*, 635 F. 2d 987 (1980), but later ignored by Weinstein, J., in favour of a 'law of national consensus' (580 F. Supp. 690, 701 (1984)).

[89] *In re 'Agent Orange', supra*, n. 74 at 821.

[90] Ibid., at 821-2.   [91] Ibid., at 828-33.

[92] *Supra*, n. 74. The idea stemmed from a student Comment, 'DES and a Proposed Theory of Enterprise Liability', 46 *Ford. L. Rev.* 63 (1978).

[93] The ruling assumed that the drug was first proved to be 'defective' in design and therefore entailed potential liability either for negligence or strict products liability.

discerned a more equitable solution in limiting each manufacturer's liability merely to its market share. That way,
when all claims had been satisfied, no one defendant would
have had to pay for more injuries than were statistically
attributable to him.

A number of objections have been raised against this
solution. Perhaps the most formidable is that it departs from
the prior art not merely by lacking all precedent but by being
incompatible with the traditional notion of tort as a system
of individual responsibility. This was not corrective but
distributive justice. Allocation of responsibility was based no
longer on proof of particular but of statistical causation.
Despite the Court's disavowal, this was indeed an industry-
wide liability. Even if defensible in terms of economic efficiency,[94] it did not conform to basic notions of individual
justice. Such a break from the whole tradition of our culture
should at best be a programme for legislation, not for judicial
reform.

Secondly, the assumption that all would work out at the
end of the day was wishful thinking.[95] The court was content
with the plaintiff's joining the manufacturers of a 'substantial
share' of the DES market, apparently viewing even 70–80 per
cent as too ambitious.[96] Moreover, it was left uncertain
whether a plaintiff could still collect the whole of his judgment
from any one defendant or was limited to the latter's
market share.[97] In the first eventuality, the cost of securing
contribution and the risk of insolvency beyond his own share
would still be borne by each defendant. How does this really
differ from solitary liability except in so far as contribution
will be regulated by reference to market share rather than
other possible criteria of responsibility?[98]

[94] See Calabresi, 'Concerning Cause and the Law of Torts', 43 *U. Chi. L.
Rev.* 69, esp. 84–91 (1973) (relation of market deterrence and cause-in-fact).

[95] See e.g. Note, 69 *Calif. L. Rev.* 1179 (1981).

[96] *Murphy* v. *E. R. Squibb & Sons*, 40 Cal. 3d 672, 221 Cal. Rptr. 447
(1985) held that 10% was insufficient, but Kaus, J., dissented on the ground
that the requirement made no sense if the liability was several only.

[97] *Brown* v. *Superior Court (Abbott Laboratories)*, 182 Cal. App. 3d 1125, 227
Cal. Rptr. 768 (1986) opted for several liability.

[98] Cf. Robinson, 'Multiple Causation in Tort Law: Reflections on the DES
Cases', 68 *Va. L. Rev.* 713 (1982).

Next is the problem of finding 'the market'. The DES market was quite fluid, with companies entering and leaving over the years; some no longer existed, others may not have any records. Is the market to be world-wide, national or regional—and, if the last, how is the area to be defined? To the extent that arbitrary judgments are made in these respects, as they must be, the relation of market share to probability is becoming ever more tenuous. Also, whenever a manufacturer who can be actually identified is held solely liable to a particular plaintiff but no allowance is made in calculating his market share, the latter will exceed his real share.

The market share theory has made slow progress elsewhere.[99] Michigan accepted 'alternative liability' on condition that all alleged tortfeasors were joined.[100] Wisconsin insisted that market share be only one factor for contribution.[101] Massachusetts held it fatal that all manufacturers were not joined and that causal exculpation was to be excluded; moreover, without entirely foreclosing the possibility of applying market-share in a suitable case, the court stressed the adverse effect it would have on development and marketing of new drugs.[102] The many lower courts, mostly federal, have split in their prediction of state law application of the theory to DES and asbestos litigation.[103] In the *Agent Orange* case, the judge considered the theory especially suitable in the context of a class action in which all claimants and manufacturers were joined and inconsistent verdicts thereby avoided.[104]

[99] See Biebel, 'DES Litigation and the Problem of Causation', 51 *Ins. Couns. J.* 223 (1984).

[100] *Abel* v. *Eli Lilly Co.*, 418 Mich. 311, 343 NW 2d 164 (1984).

[101] *Collins* v. *Eli Lilly Co.*, 116 Wis. 2d 166, 342 NW 2d 37 (1984).

[102] *Payton* v. *Abbott Labs*, 386 Mass. 540, 437 NE 2d 171 (1982). This was in answer to a question certified by the federal district court which accordingly decertified the class action: 100 FRD 336, 338-9 (D. Mass. 1983).

[103] See *In re 'Agent Orange'*, *supra*, n. 74 at 825-6.

[104] Ibid., at 823-4, 826-8. The basis of each defendant's liability was that he was, or should have been, aware of the dangerousness of the product and was therefore under a duty to warn the US government of the problems inherent in combining his product with that of the other manufacturers: at 832.

2.2   The indeterminate plaintiff

In the preceding situations the plaintiff knew that he had suffered injury as a result of another's tort but did not know precisely whose. This, the problem of the indeterminate *defendant*, has its converse image in situations where the claimant is one of several victims, only some of whom have been injured by a single tortfeasor, but who are unable to say which one among them. To illustrate this, the problem of the indeterminate *plaintiff*, suppose he is one of a group of persons exposed to a toxic emission from the defendant, but the same symptoms also emanate from independent 'background risks'. For example, in the *Nevada Nuclear Explosion* case the plaintiffs could point to a strong positive association between their cancer and exposure to ionizing radiation, but their cancer was indistinguishable from that also prevalent and attributable to unknown causes.[105] Similarly, in the *Agent Orange* case, dioxin was present in the Vietnam countryside besides the amounts in the defoliant used by the US forces, procured from seven identified American chemical companies.[106]

Typically, the association of the injury with the defendant's activity rests on statistical rather than specific (anecdotal) evidence.[107] Thus the evidence may show that, after the defendants' emission, the incidence of the particular disease rose from 100 to 190 for a given population. Here, doubts about statistical proof are compounded by the fact that it does not even tip the balance of probabilities, i.e. 50 per cent plus. In the wake of *Sindell*, proposals have been made to apply a

---

[105] *Allen* v. *US*, 588 F. Supp. 247 (D. Utah 1984). The same applies to adverse reactions from many drugs, without any clear distinction between iatrogenic and spontaneous illness: see Newdick, 'Strict Liability for Defective Drugs in the Pharmaceutical Industry', 101 *L. Q. Rev.* 405, 420–30 (1985).

[106] A similar problem has arisen in some civil rights and labour law cases where it may be impossible to identify those members of the class who suffered loss. The preferred solution is to determine the total income lost as the result of the discriminatory practice and allocate pro rata shares to qualified individuals. e.g. *Stewart* v. *General Motors Corp.*, 542 F. 2d 1043, 1055 (2nd Cir. 1976).

[107] A traditional response is Croom-Johnson, LJ's '[A] mere statistical chance will not be enough. The chance must be something lost to the individual plaintiff.' *Hotson* v. *Fitzgerald* [1987] 2 WLR 287, 304.

mirror-image solution to the instant problem so that the defendant would be held responsible for, and the plaintiffs as a group could recover, nine-nineteenths of their injuries. Professor Delgado was the first to advocate this modification of the traditional standard of proof as furthering the reputed objectives of tort liability.[108] The proposed formula would exact from the defendant an amount precisely proportioned to his share of responsibility for the total incidence of the disease in the area. Besides spreading the loss, it would promote deterrence and economic efficiency by internalizing the accident cost to the enterprise that is in the best position to reduce accidents and pass on the cost to its beneficiaries by means of insurance and price calculation.

Rather less satisfactory is the solution at the plaintiffs' end. Proportional recovery, by which each member of the class is compensated in proportion to the damages sustained by the class as a whole, undercompensates some (90 in the preceding example) and overcompensates others (100). But this is still better, so it is contended, than either to compensate none or to compensate all for the full amount of their injuries.

Class actions provide a procedural framework particularly suitable for implementing the proposed formula.[109] Indeed, unless the whole class is before the court, the formula is unworkable. In recommending this approach in the context of the class action he had certified, Judge Weinstein in the *Agent Orange* case suggested additional economies in order to avoid a tailor-made assessment of damages for each claimant.[110] As the price for making claims based on proportionality at all viable through aggregative procedures, a 'public law' approach might also dictate economies of scale in fixing benefits. Thus if the class were for one type of injury, a compensation schedule to calculate average losses could be developed by sampling techniques. Better still, claims could be paid on a 'fixed and somewhat arbitrary schedule' as under bureaucratic compensation programmes. Perhaps most

---

[108] Delgado, 'Beyond *Sindell*: Relaxation of Cause-in-Fact Rules for Indeterminate Plaintiffs', 70 *Calif. L. Rev.* 881 (1982); Rosenberg, *supra*.

[109] *In re 'Agent Orange'*, *supra*, n. 74, at 837–9; Rosenberg, *supra*.

[110] Ibid., at 838–9; adopting Rosenberg's thesis, *supra*, at 916–24.

important from a practical point of view, the class action context strongly encourages settlements on an agreed basis. On the other hand, the departure from traditional concepts, propounded in *Agent Orange*, is manifold and startling. On the basis of mere statistical evidence of a product's propensity for injury, it sanctions a cause of action by unidentified plaintiffs against unidentified defendants without specific proof of the defective nature of the product or of its having caused injury to a particular plaintiff. In short, most elements of products liability have been collapsed into mere statistical proof of causation.[111] While it is true that, strictly speaking, the court's reasoning related only to the fairness of the settlement, it sought approval for an approach to liability that would sever most links to traditional tort principles.

3. POSTSCRIPT

The prospect of enlisting class actions for the radical solution of social problems envisaged by these proposals enjoys far from universal support. For even if the goals are worthy, to entrust such drastic legal change to a selection of activist judges instead of to the traditional venue for political decision in a democracy challenges accepted constitutional understandings. The tendency towards collectivist solutions among so-called 'forward-looking' or 'progressive' courts has moved tort law increasingly away from the traditional perceptions of 'corrective justice'. The displacement of individualized by aggregative procedures hastens this trend to 'distributive justice'.[112] The distance separating these perspectives is nowhere more strikingly illustrated than by the justification for the *Agent Orange* settlement enforced by Judge Weinstein:

---

[111] Sherman, 'Agent Orange and the Problem of the Indeterminate Plaintiff', 52 *Brooklyn L. Rev.* 369, 390 (1986). Significantly, Judge Weinstein later dismissed the claims of 'opt-outs' for failure of causation: 611 F. Supp. 1223 (EDNY 1985). For comprehensive analysis of Agent Orange litigation see Schuck, *supra*.

[112] See Glenn, 'Class Actions and the Theory of Tort and Delict', 35 *U. Tor. L. J. 287 (1985)*.

Based on all the information presently available, the procedural posture of the litigation, the difficulty any plaintiff would have in establishing a case against any one or more of the defendants, the uncertainties associated with a trial, and the unacceptable burdens on the plaintiffs' and defendants' legal staffs and the courts, the proposed settlement appears to be reasonable. It appears to be in the public's as well as the parties' interest.

Even though the evidence presented to the court to date suggests that the case is without merit, many of those who testified at the Fairness Hearing indicated that they sought its prosecution not for money but for public vindication. . . .

The publicity attendant upon the settlement and subsequent Fairness Hearings, where the court heard almost 500 witnesses from all parts of the nation, undoubtedly did serve once again to bring to the public's attention how unfairly Vietnam veterans have been treated. They have been abused, rejected and humiliated after serving bravely. Their voices should be heeded by the government and public for whom they fought. . . .

Whether or not that pain was caused by Agent Orange, it is shared by a disproportionately large number of Vietnam veterans. They and their families should receive recognition, medical treatment and financial support. . . . The public received the 'benefit' of combat service and should help defray the cost. . . .

Our country is rich in public and private resources of every kind. Those resources should be made available to members of the class.[113]

This statement not only turns its back on any dichotomy between principle and policy in judicial decision-making,[114] but invokes a goal of social psychology as justification for wealth redistribution, far removed from any conventional objective of accident compensation policy, let alone of tort law.

Mass litigation is not the only solution for mass accidents. Whether the procedure is individualized or aggregative, the tort system reveals its inefficiencies in starkest colour in dealing with mass claims. The funds available for compensation are limited by the resources of the defendants, including liability

[113] *In re 'Agent Orange', supra*, n. 74 at 857-8. The settlement was upheld on appeal: 818 F.2d 145 (2nd Cir. 1987).

[114] The distinction, resuscitated by Dworkin's *Taking Rights Seriously* (1977), attained further prominence in England through the exchange between Lords Scarman and Edmund-Davies in *McLoughlin* v. *O'Brian*, [1983] 1 AC 410.

insurance cover. That even industrial giants can be driven into bankruptcy has been translated from rhetoric to reality in the wake of the asbestos litigation. The most depressing feature is that the exorbitant cost of administering the tort system not only threatens the survival of industries peculiarly exposed to the risk of mass claims, like the pharmaceutical and chemical industries, but depletes the available funds for compensating victims by staggering litigation costs. These, it must be remembered, include not only lawyers' fees of successful claimants but also the cost to defendants of successfully resisting claims—a factor that has been most influential in persuading defendants to settle rather than continuing a spree of pyrrhic victories.

As the preceding discussion revealed, even the substantive tort law is unequal to dealing fairly and effectively with systemic problems of causation in mass accident and mass exposure cases. To the extent that traditional rules are already being modified in order to facilitate recovery by victims, the tort system is being distorted, even superseded. If the conventional tort law is thus proving itself inadequate to the task, should we not, instead of merely tinkering with it, consider the more radical solution of entirely replacing it by a compensation scheme?[115] This would have several advantages: first, it would eliminate most contestable issues that are presently both the cause and justification of the high legal costs. By assuring compensation to victims on a no-fault basis, the contingency factor in the claimants' representation would disappear entirely and the occasion for protracted adversary proceedings would be substantially reduced. Secondly, victims could count on certain and speedier compensation. Under the present system, defendants who are minded to 'play tough' could jam the courts for decades and wear down the patience of claimants and their attorneys to the point of attrition. Thirdly, the cost of compensation can be rationally internalized by making advance provisions for a

---

[115] See Sugarman, *supra*, especially at 596–603, where the author illustrates his thesis by reference to Agent Orange, Bendectin, Asbestos and IUD litigation. Also Stapleton, *Disease and the Compensation Debate* (1986), advocating non-cause related compensation for disease (traumatic and other).

compensation fund, contributed to by a particular industry and ultimately borne by the consumer public, which is often identical with those exposed to the risk. It would be a legislative decision to prescribe the magnitude of the fund and compensation tariffs. By standardizing benefits, as is already advocated for some class actions, further economies can be achieved in administering the scheme. In any event, important industries whose social utility obviously outweighs the risk their activities portend to public safety would no longer be faced with the ever present threat of extinction. By internalizing the compensation cost to the risk-creating enterprise or industry, some 'general deterrence' can be preserved in the interest of accident prevention.[116] Moreover, some of the industries peculiarly affected, like pharmaceuticals, are already under stringent regulatory supervision, thus providing added safeguard against abuse.

A number of countries have already replaced the tort system by compensation funds for drug injuries.[117] Nuclear legislation in many others, including the United States, has set up a structure for minimum insurance and an upper limit to liability exposure based on a no-fault principle.[118] This trend may be a harbinger for the eventual disappearance of the tort system for all personal injury by accident, as has already occurred in New Zealand.[119]

[116] See Calabresi, *The Cost of Accidents: A Legal and Economic Analysis* (1970) for his thesis on 'general deterrence'.
[117] See Fleming, 'Drug Injury Compensation Plans', 30 *Am. J. Comp. L.* 297 (1982).
[118] (US) Atomic Energy Act, 42 USC § 2210 (the Price–Anderson Act).
[119] Accident Compensation Act 1974. See Palmer, *Compensation for Incapacity* (1979); Ison, *Accident Compensation* (1980); Blair, *Accident Compensation in New Zealand* (1978).

INDEX